Praise for Larry Hatcher

The problems are interesting, and the tasks required are those a researcher must undertake. Students who work through all of the exercises in this book will definitely be confident and prepared to use SAS to conduct these analyses independently.

Sheri Bauman, Ph.D.
Assistant Professor
Department of Educational Psychology
University of Arizona, Tucson

My graduate students and I have had a good chance to look over the new workbook...The general consensus is that it is extremely successful on a number of levels. [Larry Hatcher has] thought of everything...We especially like how [he] incorporated additional explanations to help students navigate the computer itself.

Frank Pajares
Winship Distinguished Research Professor
Emory University

SAS Publishing

EXERCISES

Step-by-Step BASIC STATISTICS Using SAS®

LARRY HATCHER, PH.D.

The Power to Know®

The correct bibliographic citation for this manual is as follows: Hatcher, Larry. 2003. *Step-by-Step Basic Statistics Using SAS®: Exercises,* Cary, NC: SAS Institute Inc.

Step-by-Step Basic Statistics Using SAS®: Exercises

ISBN 1-59047-149-0

SAS Institute Inc., SAS Campus Drive, Cary, North Carolina 27513.

1st printing, April 2003

Note that text corrections may have been made at each printing.

Contents

Acknowledgments

During the development of these books, Caroline Brickley, Gretchen Rorie Harwood, Stephenie Joyner, Sue Kocher, Patsy Poole, and Hanna Schoenrock served as editors. All were positive, supportive, and helpful. They made the books stronger, and I thank them for their guidance.

A number of other people at SAS made valuable contributions in a variety of areas. My sincere thanks go to those who reviewed the books for technical accuracy and readability: Jim Ashton, Jim Ford, Marty Hultgren, Catherine Lumsden, Elizabeth Maldonado, Paul Marovich, Ted Meleky, Annette Sanders, Kevin Scott, Ron Statt, and Morris Vaughan. I also thank Candy Farrell and Karen Perkins for production and design; Joan Stout for indexing; Cindy Puryear and Patricia Spain for marketing; and Cate Parrish for the cover designs.

Special thanks to my wife Ellen, who was loving and supportive throughout.

EXERCISES

Overview

This section contains two exercises for each chapter, beginning with Chapter 3, "Writing and Submitting SAS Programs." Each exercise includes an overview that describes the purpose of the exercise, a detailed description of the assignment, a list of the items to be handed in, and hints on completing the assignment.

Exercises for Chapter 3: Writing and Submitting SAS Programs

EXERCISES 3

Exercise 3.1: Computing Mean Height, Weight, and Age

Overview

In this exercise, you will use the SAS windowing environment to write and submit a simple SAS program. You will print the SAS program, along with the resulting SAS log and SAS output files.

Your Assignment

1. First, you should be sure that all of your windows are clear (that is, you do not have a program in the Editor window, and you do not have output in the Log or Output windows). If you have just opened SAS, all of your windows should be clear. However, if you have already created and submitted SAS programs during this session, then you will have to clear your windows. If this is the case, then select the following (Remember that the Window menu might contain the name that you gave to your SAS program, rather than the word EDITOR, which appears here.):

   ➔ **Window** ➔ **Editor** ➔ **Edit** ➔ **Clear All**

   ➔ **Window** ➔ **Log** ➔ **Edit** ➔ **Clear All**

   ➔ **Window** ➔ **Output** ➔ **Edit** ➔ **Clear All**

All three of your windows should now be clear. If it is not already, make sure that the Editor window is the active window. Select the following:

Window ➔ Editor

2. Use the Editor window to write the following program (see the notes following the program before you begin):

```
1     OPTIONS  LS=80  PS=60;
2     DATA D1;
3        INPUT  SUB_NUM
4               HEIGHT
5               WEIGHT
6               AGE ;
7     DATALINES;
8     1 64 140 20
9     2 68 170 28
10    3 74 210 20
11    4 60 110 32
12    5 64 130 22
13    6 68 170 23
14    7 65 140 22
15    8 65 140 22
16    9 68 160 22
17    ;
18    PROC MEANS DATA=D1;
19       VAR  HEIGHT  WEIGHT  AGE;
20       TITLE1 'type your name here';
21    RUN;
```

Some notes regarding this program:

- Do not type the line numbers that appear on the left (i.e., *1*, *2*, *3*, and so on); these line numbers are provided simply for you to use as a guide.
- On line 3 in the INPUT statement, the first variable name is *SUB_NUM*. Be sure to type this variable name using an underscore (_) and not a hyphen (-).
- On line 20 in the TITLE1 statement, be sure to type your first and last name enclosed in single quotation marks.
- Don't worry if you don't understand the data that are contained in this program or the analysis that it performs. You are only working with this

program at this time in order to gain experience with using the SAS windowing environment.

3. Submit the program, and, if necessary, correct it so that it runs without errors.
4. Review your log and output windows for possible errors.
5. When you are sure that your program ran correctly without errors, print your SAS program, SAS log, and SAS output.

What You Will Hand In

Hand in the following materials stapled together in this order:

1. A printout of your SAS program. At the top of this printout, write "Exercise 3.1" and your first and last names.
2. A printout of your SAS log.
3. A printout of your SAS output. Your SAS output must have your name on it in the title position in order to receive full credit. If you typed your name in the TITLE1 statement of your SAS program correctly, it should appear in the title position on the output.

Hint

If your program ran correctly, your output should resemble the following output (except that your name will appear where JANE DOE now appears):

```
                              JANE DOE                                 1
                         The MEANS Procedure

Variable    N          Mean       Std Dev       Minimum       Maximum
 ----------------------------------------------------------------------
HEIGHT      9    66.2222222     3.8980052    60.0000000    74.0000000
WEIGHT      9   152.2222222    29.0593263   110.0000000   210.0000000
AGE         9    23.4444444     3.9721251    20.0000000    32.0000000
 ----------------------------------------------------------------------
```

Exercise 3.2: Computing Mean Age, IQ, and Income

Overview

In this exercise, you will again use the SAS windowing environment to write and submit a simple SAS program. You will print the SAS program, along with the resulting SAS log and SAS output files.

Your Assignment

1. First, you should be sure that all of your windows are clear (that is, you do not have a program in the Editor window, and you do not have output in the Log or Output windows). If you have just opened SAS, all of your windows should be clear. However, if you have already created and submitted SAS programs during this session, then you will have to clear your windows. If this is the case, then select the following (Remember that the Window menu might contain the name that you gave to your SAS program, rather than the word EDITOR, which appears here.):

   ➔ **Window** ➔ **Editor** ➔ **Edit** ➔ **Clear All**

   ➔ **Window** ➔ **Log** ➔ **Edit** ➔ **Clear All**

   ➔ **Window** ➔ **Output** ➔ **Edit** ➔ **Clear All**

   All three of your windows should now be clear. If it is not already, make sure that the Editor window is the active window. Select the following:

   Window ➔ **Editor**

2. Use the Editor window to write the following program (see the notes following the program before you begin):

```
1       OPTIONS  LS=80  PS=60;
2       DATA D1;
3          INPUT  AGE
4                 IQ
5                 INCOME;
6       DATALINES;
7       36 110 25000
8       28 105 29000
9       55  99 58000
10      44 102 45000
11      46 108 51000
12      39  90 46000
13      29 100 31000
14      40  92 43000
15      32 105 31000
16      ;
17      PROC MEANS DATA=D1;
18         VAR  AGE  IQ  INCOME;
19         TITLE1 ' type your name here ';
20      RUN;
```

Some notes regarding this program:

- Do not type the line numbers that appear on the left (i.e., *1*, *2*, *3*, and so on); these line numbers are provided simply for you to use as a guide.
- On line 19 in the TITLE1 statement, be sure to type your first and last names enclosed in single quotation marks.
- Don't worry if you don't understand the data that are contained in this program or the analysis that it performs. You are only working with this program at this time in order to gain experience with using the SAS windowing environment.

3. Submit the program, and, if necessary, correct it so that it runs without errors.

4. Review your log and output windows for possible errors.

5. When you are sure that your program ran correctly without errors, print your SAS program, SAS log, and SAS output.

What You Will Hand In

Hand in the following materials stapled together in this order:

1. A printout of your SAS program. At the top of this printout, write "Exercise 3.2" and your first and last names.
2. A printout of your SAS log.
3. A printout of your SAS output. Your SAS output must have your name on it in the title position in order to receive full credit. If you typed your name in the TITLE1 statement of your SAS program correctly, it should appear in the title position on the output.

Hint

If your program ran correctly, your output should resemble the following output (except that your name will appear where JOHN DOE now appears):

```
                              JOHN DOE                                1

                       The MEANS Procedure

Variable  N          Mean       Std Dev       Minimum        Maximum
--------------------------------------------------------------------
AGE       9    38.7777778     8.7289429    28.0000000     55.0000000
IQ        9   101.2222222     6.7966495    90.0000000    110.0000000
INCOME    9      39888.89      11307.57      25000.00       58000.00
--------------------------------------------------------------------
```

EXERCISES

4

Exercises for Chapter 4: Data Input

Exercise 4.1: Creating and Analyzing a Data Set Containing LAT Test Scores

Overview

In this exercise, you will create a small data set and analyze it using PROC MEANS, PROC FREQ, and PROC PRINT. The purpose of the exercise is to give you experience in organizing data, creating valid SAS variable names, using SAS procedures, and, if necessary, debugging programs that contain errors.

The Study

Suppose that you are conducting research on a fictitious instrument called the Learning Aptitude Test (LAT). Assume that the LAT is a standardized test that is administered to people (subjects) who plan to go to college, and its scores are used to make admissions decisions. The LAT consists of two subtests: the LAT Verbal subtest, and the LAT Math subtest.

From a sample study, you obtain scores for nine subjects based on data from the following six variables:

- **LAT Verbal subtest.** Possible scores can range from 200-800.
- **LAT Math subtest.** Possible scores can range from 200-800.
- **Subject sex.** Possible values are "F" for females and "M" for males.

- **Scores on mid-term test #1**, which was given in a general psychology course. Possible scores can range from 0-99.
- **Scores on mid-term test #2**, which was given in a general psychology course. Possible scores can range from 0-99.
- **Scores on mid-term test #3**, which was given in a general psychology course. Possible scores can range from 0-99.

Data Set to Be Analyzed

Table 4.E1.1 provides data for the nine subjects using the previously discussed six variables. The Subject number column in the table lists a unique number for each subject.

Table 4.E1.1

Data from the LAT Study

				Scores on mid-term tests		
Subject number	LAT Verbal	LAT Math	Sex	#1	#2	#3
01	510	520	F	89	92	92
02	530	.	M	88	75	89
03	620	600	F	95	90	88
04	.	490	F	80	.	78
05	650	600	M	97	95	96
06	550	510	F	76	70	78
07	420	480	.	88	81	85
08	400	410	M	90	88	88
09	590	610	F	90	92	95

Your Assignment

1. Write a SAS program that will input and analyze the preceding data set. When you write the DATA step of this program, do the following:
   - Use the SAS programs presented in Chapter 4, "Data Input" of the *Student Guide* as models.
   - Type all of the information from Table 4.E1.1. You should type the subject number variable that appears as the first column in the table. Give this variable the SAS variable name *SUB_NUM*. When you type this SAS variable name, make sure you use an underscore (_) and not a hyphen (-). When you type subject numbers such as 01, make sure you type a zero (0) and not the letter O.
   - Create SAS variable names for the other variables in the table. Make sure that the SAS variable names you create are meaningful and conform to the rules for SAS variable names presented in Chapter 4 of the *Student Guide*. Also, remember not to use hyphens in your SAS variable names.
   - The periods that appear in the preceding table represent missing data. Be sure that you type these periods in your SAS program as well.
2. When you write the PROC step of this program, do the following:
   - Use PROC MEANS to analyze the following numeric variables: LAT Verbal scores, LAT Math scores, mid-term test #1, mid-term test #2, and mid-term test #3.
   - Insert a TITLE1 statement after this PROC MEANS statement. Type your full name in the TITLE1 statement so that it will appear on the output.
   - Use PROC FREQ to analyze the sex variable.
   - Use PROC PRINT to create a printout of the raw data. Omit the VAR statement so that it will print out the raw data for all variables in the data set.
3. Submit the program, and if necessary, correct it so that it runs without errors.
4. Review your log window for possible errors.

5. Review your output window for possible errors. First, review the results of PROC MEANS and PROC FREQ for errors in the SAS program, as described in Chapter 4, "Interpreting the Results Produced by PROC MEANS" and "Interpreting the Results Produced by PROC FREQ." Next, check the results of PROC PRINT against Table 4.E1.1 to verify that you typed your data correctly.

What You Will Hand In

Hand in the following materials stapled together in this order:

1. A printout of your SAS program.
2. A printout of your SAS log.
3. A printout of your SAS output.

Hints

If your SAS program ran correctly, your output should resemble the following output (except that your name will appear where "JOHN DOE" now appears). You do not necessarily have to use the SAS variable names that appear in this output.

```
                          JOHN DOE                                  1

                    The MEANS Procedure

Variable  N          Mean       Std Dev       Minimum       Maximum
-------------------------------------------------------------------
LAT_V     8   533.7500000    89.2728562   400.0000000   650.0000000
LAT_M     8   527.5000000    70.8620390   410.0000000   610.0000000
TEST_1    9    88.1111111     6.5849154    76.0000000    97.0000000
TEST_2    8    85.3750000     9.0386077    70.0000000    95.0000000
TEST_3    9    87.6666667     6.5000000    78.0000000    96.0000000
-------------------------------------------------------------------
```

JOHN DOE 2

The FREQ Procedure

SEX	Frequency	Percent	Cumulative Frequency	Cumulative Percent
F	5	62.50	5	62.50
M	3	37.50	8	100.00

Frequency Missing = 1

JOHN DOE 3

Obs	SUB_NUM	LAT_V	LAT_M	SEX	TEST_1	TEST_2	TEST_3
1	1	510	520	F	89	92	92
2	2	530	.	M	88	75	89
3	3	620	600	F	95	90	88
4	4	.	490	F	80	.	78
5	5	650	600	M	97	95	96
6	6	550	510	F	76	70	78
7	7	420	480		88	81	85
8	8	400	410	M	90	88	88
9	9	590	610	F	90	92	95

Exercise 4.2: Creating and Analyzing a Data Set Containing Information About Volunteerism

Overview

In this exercise, you will create a small data set and analyze it using PROC MEANS, PROC FREQ, and PROC PRINT. The purpose of the exercise is to give you experience in organizing data, creating valid SAS variable names, using SAS procedures, and, if necessary, debugging programs that contain errors.

The Study

Suppose that you are a sociologist conducting research on volunteerism. You have interviewed 11 people (subjects) and have asked them if they have served as a volunteer in their community during the past year. You also asked them about their income, and obtained some demographic information (e.g., age, sex).

Now you need to obtain some descriptive statistics for your data. You will compute means for the numeric variables, and you will create frequency tables for the character variables. You will also print out the raw data set.

From the study, you obtain data for 11 subjects based on the following six variables:

- **Subject number.** Represents a unique subject number that is assigned to each subject.
- **Age.** Represents each subject's age in years.
- **Sex.** Represents each subject's sex. Here, "F" represents female subjects and "M" represents male subjects.
- **Race.** Represents each subject's race. Here, "AF" represents African Americans, "AS" represents Asian Americans, and "CA" represents Caucasians.
- **Income.** Represents each subject's annual income in dollars.

- **Served as volunteer?** Indicates whether the subject has served as a volunteer during the past year. "Y" means yes, and "N" means no.

Data Set to Be Analyzed

Table 4.E2.1 provides data for the 11 subjects.

Table 4.E2.1

Data from the Volunteerism Study

Subject number	Age	Sex	Race	Income	Served as volunteer?
01	20	F	AF	75000	Y
02	32	F	AF	.	Y
03	34	M	CA	26000	Y
04	.	M	CA	31000	N
05	29	F	CA	23000	Y
06	41	F	AS	29000	Y
07	58	.	AS	59000	N
08	37	.	CA	28000	N
09	25	M	CA	1000	Y
10	39	M	AF	55000	Y
11	44	F	CA	50000	N

Your Assignment

1. Write a SAS program that will input and analyze the preceding data set. When you write the DATA step of this program, do the following:
   - Use the SAS programs presented in Chapter 4, "Data Input" of the *Student Guide* as models.
   - Type all of the information from Table 4.E2.1. You should type the subject number variable that appears as the first column in the table. Give this variable the SAS variable name SUB_NUM. When you type this SAS variable name, make sure you use an underscore (_) and not a

hyphen (-). When you type subject numbers such as 01, make sure you type a zero (0) and not the letter O.

- Create SAS variable names for the other variables in the table. Make sure that the SAS variable names you create are meaningful and conform to the rules for SAS variable names presented in Chapter 4 of the *Student Guide*. Also, remember not to use hyphens in these SAS variable names.
- The periods in the preceding table represent missing data. Be sure to type these periods in your SAS data set.

2. When you write the PROC step of this program, do the following:
   - Use PROC MEANS to analyze the following numeric variables: Subject number, age, and income.
   - Insert a TITLE1 statement after this PROC MEANS statement. Type your full name in the TITLE1 statement so that it will appear on the output.
   - Use PROC FREQ to analyze the following character variables: sex, race, and the "served as a volunteer?" variable.
   - Use PROC PRINT to create a printout of the raw data. Omit the VAR statement so that it will print out the raw data for all variables in the data set.
3. Submit the program, and, if necessary, correct it so that it runs without errors.
4. Review your log window for possible errors.
5. Review your output window for possible errors. First, review the results of PROC MEANS and PROC FREQ for errors in the SAS program, as described in Chapter 4, "Interpreting the Results Produced by PROC MEANS" and "Interpreting the Results Produced by PROC FREQ." Next, check the results of PROC PRINT against Table 4.E2.1 to verify that you typed your data correctly.

What You Will Hand In

Hand in the following materials stapled together in this order:

1. A printout of your SAS program.
2. A printout of your SAS log.
3. A printout of your SAS output.

Hints

If your SAS program ran correctly, your output should resemble the following output (except that your name will appear where "JOHN DOE" now appears). You do not necessarily have to use the SAS variable names that appear in this output.

```
                              JOHN DOE                                   1

                        The MEANS Procedure

Variable    N          Mean        Std Dev        Minimum        Maximum
---------------------------------------------------------------------------
SUB_NUM    11     6.0000000      3.3166248      1.0000000     11.0000000
AGE        10    35.9000000     10.6921570     20.0000000     58.0000000
INCOME     10      37700.00       21628.43        1000.00       75000.00
---------------------------------------------------------------------------
```

```
                         JOHN DOE                              2

                   The FREQ Procedure

                                      Cumulative    Cumulative
     SEX    Frequency     Percent     Frequency      Percent
     ---------------------------------------------------------
     F             5       55.56             5        55.56
     M             4       44.44             9       100.00

                Frequency Missing = 2




                                      Cumulative    Cumulative
    RACE    Frequency     Percent     Frequency      Percent
    ----------------------------------------------------------
    AF             3       27.27             3        27.27
    AS             2       18.18             5        45.45
    CA             6       54.55            11       100.00




                                       Cumulative    Cumulative
   VOLUNT    Frequency     Percent     Frequency      Percent
   -------------------------------------------------------------
   N                 4       36.36             4        36.36
   Y                 7       63.64            11       100.00
```

JOHN DOE 3

Obs	SUB_NUM	AGE	SEX	RACE	INCOME	VOLUNT
1	1	20	F	AF	75000	Y
2	2	32	F	AF	.	Y
3	3	34	M	CA	26000	Y
4	4	.	M	CA	31000	N
5	5	29	F	CA	23000	Y
6	6	41	F	AS	29000	Y
7	7	58		AS	59000	N
8	8	37		CA	28000	N
9	9	25	M	CA	1000	Y
10	10	39	M	AF	55000	Y
11	11	44	F	CA	50000	N

EXERCISES 5

Exercises for Chapter 5: Creating Frequency Tables

Exercise 5.1: Using PROC FREQ to Analyze LAT Data

Overview

In this exercise, you will create a SAS data set that contains scores from a fictitious test called the Learning Aptitude Test (LAT). You will use PROC FREQ to create a frequency table for one of the variables. You will then answer a series of questions about the information presented in the frequency table (questions about frequencies, percentages, and so forth). When you respond to these questions, write your answers on a separate sheet of paper, and circle the appropriate numbers on your SAS output.

The Study

Suppose that you are conducting research using a fictitious instrument called the Learning Aptitude Test (LAT). The LAT is a standardized test that is administered to people (subjects) who plan to go to college, and its scores are used to make admissions decisions. The LAT consists of three subtests: the LAT Verbal subtest, the LAT Math subtest, and the LAT Analytical subtest.

From a sample study, you obtain scores for 18 subjects based on data from the following four variables:

- **LAT Verbal subtest.** Possible scores can range from 200-800.

- **LAT Math subtest.** Possible scores can range from 200-800.
- **LAT Analytical subtest.** Possible scores can range from 200-800.
- **Area that the subject is majoring in at college.** You use the value "A" to identify subjects majoring in the College of Arts & Sciences, "B" to identify those majoring in the College of Business, and "E" to identify those majoring in the College of Education.

Data Set to be Analyzed

Table 5.E1.1 provides data for the 18 subjects using the previously discussed four variables. The Subject number column in the table lists a unique number for each subject.

Table 5.E1.1

Data from the LAT Study

Subject number	LAT-Verbal	LAT-Math	LAT-Analytical	Major
01	540	540	540	A
02	510	560	550	A
03	500	520	530	A
04	490	550	530	A
05	520	510	510	A
06	520	500	530	A
07	510	470	520	E
08	530	490	520	B
09	500	460	480	B
10	490	480	500	E
11	510	470	470	B
12	500	450	490	E
13	500	460	540	E
14	480	440	490	B
15	490	430	460	B
16	480	450	480	E
17	470	440	470	E
18	460	450	480	B

Your Assignment Regarding the SAS Program

1. Write a SAS program that will input and analyze the preceding data. When you write this program, do the following:
   - Use the DATA step from the SAS program that was presented in Chapter 5, "Creating Frequency Tables" of the *Student Guide* as a model. If necessary, you can also refer to Chapter 4: Data Input for more information.
   - Type all of the information from Table 5.E1.1. You should type the Subject number variable as the first column as it appears in Table 5.E1.1. When you type subject numbers such as 01, make sure that you type a zero (0) and not the letter O.
   - Assign the following SAS variable names to the variables. For the first four variable names, make sure that you use an underscore (_) and not a hyphen (-).
     - Use the SAS variable name SUB_NUM to represent Subject numbers (the first column in Table 5.E1.1).
     - Use the SAS variable name LAT_V for the LAT Verbal subtest.
     - Use the SAS variable name LAT_M for the LAT Math subtest.
     - Use the SAS variable name LAT_A for the LAT Analytical subtest.
     - Use the SAS variable name MAJOR for the student's major.

2. After you write the DATA step, do the following:
   - Add statements that perform PROC FREQ on the variable LAT_V (LAT Verbal subtest). That is, use PROC FREQ to create a simple frequency table for LAT_V. If you need more information, see "Using PROC FREQ to Create a Frequency Table" from the *Student Guide*.
   - Type your full name in the TITLE1 statement so that it will appear in the output.

3. Submit the program, and, if necessary, correct it so that it runs without errors.

4. Review your log window and output files for possible errors.

Your Assignment Regarding the SAS Output

In this exercise, you will answer a series of questions about the frequency table that was produced by your SAS program. Write your answers to these questions on BOTH of the following:

- on a blank sheet of paper. At the top of the sheet write (a) your full name, and (b) the title "Exercise 5.1: Answers to Questions." Number the first seven lines on this sheet with the numbers 1 through 7 (you will write your answers to the following seven questions on these lines).
- on the SAS output page that was produced by PROC FREQ (the frequency table for LAT_V). On this output page, you will circle numbers to represent your answers.

Questions Regarding the SAS Output

1. **Question**: What is the lowest observed value for the LAT_V variable?
   - Write the answer on your sheet of paper that is titled "Exercise 5.1: Answers to Questions."
   - In the frequency table that was generated by PROC FREQ on your SAS output page, circle the number that constitutes your answer. Write "Q1" next to this circled answer.
2. **Question**: What is the highest observed value for the LAT_V variable?
   - Write the answer on your sheet of paper that is titled "Exercise 5.1: Answers to Questions."
   - In the frequency table that was generated by PROC FREQ on your SAS output page, circle the number that constitutes your answer. Write "Q2" next to this circled answer.
3. **Question**: How many people had an LAT Verbal score of 490? (That is, what is the frequency for people who displayed a score of 490 on the LAT_V variable?)
   - Write the answer on your sheet of paper that is titled "Exercise 5.1: Answers to Questions."

- In the frequency table that was generated by PROC FREQ on your SAS output page, circle the number that constitutes your answer. Write "Q3" next to this circled answer.

4. **Question**: What percentage of people had an LAT Verbal score of 500?
   - Write the answer on your sheet of paper that is titled "Exercise 5.1: Answers to Questions."
   - In the frequency table that was generated by PROC FREQ on your SAS output page, circle the number that constitutes your answer. Write "Q4" next to this circled answer.

5. **Question**: How many people had an LAT Verbal score of 480 or lower?
   - Write the answer on your sheet of paper that is titled "Exercise 5.1: Answers to Questions."
   - In the frequency table that was generated by PROC FREQ on your SAS output page, circle the number that constitutes your answer. Write "Q5" next to this circled answer.

6. **Question**: What percentage of people had an LAT Verbal score of 520 or lower?
   - Write the answer on your sheet of paper that is titled "Exercise 5.1: Answers to Questions."
   - In the frequency table that was generated by PROC FREQ on your SAS output page, circle the number that constitutes your answer. Write "Q6" next to this circled answer.

7. **Question**: What is the total number of usable observations for the LAT_V variable in this data set?
   - Write the answer on your sheet of paper that is titled "Exercise 5.1: Answers to Questions."
   - In the frequency table that was generated by PROC FREQ on your SAS output page, circle the number that constitutes your answer. Write "Q7" next to this circled answer.

What You Will Hand In

Hand in the following materials stapled together in this order:

1. A printout of your SAS program.
2. A printout of your SAS log.
3. A printout of your SAS output (from PROC FREQ), with the appropriate numbers circled according to the preceding directions.
4. Your sheet of paper titled "Exercise 5.1: Answers to Questions," on which you wrote answers to the preceding questions.

Hint

If your SAS program ran correctly, your output should resemble the following:

```
                         JOHN DOE                              1

                    The FREQ Procedure

                                         Cumulative    Cumulative
LAT_V    Frequency       Percent          Frequency      Percent
------------------------------------------------------------------
 460             1          5.56                  1          5.56
 470             1          5.56                  2         11.11
 480             2         11.11                  4         22.22
 490             3         16.67                  7         38.89
 500             4         22.22                 11         61.11
 510             3         16.67                 14         77.78
 520             2         11.11                 16         88.89
 530             1          5.56                 17         94.44
 540             1          5.56                 18        100.00
```

Exercise 5.2: Using PROC FREQ to Analyze Exercise Data

Overview

In this exercise, you will create a SAS data set that contains scores from a fictitious study about exercising behavior. You will use PROC FREQ to create a frequency table for one of the variables. You will then answer a series of questions about the information presented in the frequency table (questions about frequencies, percentages, and so forth). When you respond to these questions, write your answers on a separate sheet of paper, and circle the appropriate numbers on your SAS output.

The Study

Suppose that you are conducting research about exercise, and you want to obtain some basic information about how many minutes the typical college student exercises during the week. You select a random sample of 19 college students (subjects), and you ask them to complete an anonymous questionnaire in which they indicate how many minutes they spend exercising during the typical week.

Data Set to be Analyzed

Table 5.E2.1 provides data for the 19 subjects. The Subject number column in the table lists a unique number for each subject. The Minutes exercising column lists the number of minutes that each subject spends exercising during the typical week.

Table 5.E2.1

Data from the Exercise Study

Subject number	Minutes exercising
01	120
02	20
03	300
04	0
05	0
06	60
07	0
08	240
09	40
10	0
11	20
12	20
13	0
14	180
15	60
16	0
17	30
18	20
19	180

Your Assignment Regarding the SAS Program

1. Write a SAS program that will input and analyze the preceding data set. When you write this program, do the following:
   - Use the DATA step from the SAS program that was presented in Chapter 5, "Creating Frequency Tables" of the *Student Guide* as a model. If necessary, you can also refer to Chapter 4: Data Input for more information.
   - Type both of the variables from Table 5.E2.1. You should type the Subject number variable as the first column as it appears in Table 5.E2.1. When you type subject numbers such as 01, make sure that you type a zero (0) and not the letter O.

- Assign the following SAS variable names to the variables:
  - Use the SAS variable name SUB_NUM to represent Subject numbers (the first column in Table 5.E2.1). For this variable name, make sure that you use an underscore (_) and not a hyphen (-).
  - Use the SAS variable name MINUTES for the Minutes exercising variable.

2. After you write the DATA step, do the following:
   - Add statements that perform PROC FREQ on the variable MINUTES. That is, use PROC FREQ to create a simple frequency table for MINUTES. If you need more information, see "Using PROC FREQ to Create a Frequency Table" from the *Student Guide*.
   - Type your full name in the TITLE1 statement so that it will appear in the output.
3. Submit the program, and, if necessary, correct it so that it runs without errors.
4. Review your log window and output files for possible errors.

Your Assignment Regarding the SAS Output

In this exercise, you will answer a series of questions about the frequency table that was produced by your SAS program. Write your answers to these questions on BOTH of the following:

- on a blank sheet of paper. At the top of the sheet write (a) your full name, and (b) the title "Exercise 5.2: Answers to Questions." Number the first seven lines on this sheet with the numbers 1 through 7 (you will write your answers to the following seven questions on these lines).
- on the SAS output page that was produced by PROC FREQ (the frequency table for LAT_V). On this output page, you will circle numbers to represent your answers.

Questions Regarding the SAS Output

1. **Question**: What is the lowest observed value for the MINUTES variable?
   - Write the answer on your sheet of paper that is titled "Exercise 5.2: Answers to Questions."
   - In the frequency table that was generated by PROC FREQ on your SAS output page, circle the number that constitutes your answer. Write "Q1" next to this circled answer.

2. **Question**: What is the highest observed value for the MINUTES variable?
   - Write the answer on your sheet of paper that is titled "Exercise 5.2: Answers to Questions."
   - In the frequency table that was generated by PROC FREQ on your SAS output page, circle the number that constitutes your answer. Write "Q2" next to this circled answer.

3. **Question**: How many people indicated that they exercise for exactly 180 minutes (no more and no less) in a typical week? (In other words, what is the frequency for people who displayed a score of 180 on MINUTES?)
   - Write the answer on your sheet of paper that is titled "Exercise 5.2: Answers to Questions."
   - In the frequency table that was generated by PROC FREQ on your SAS output page, circle the number that constitutes your answer. Write "Q3" next to this circled answer.

4. **Question**: What percentage of people indicated that they exercise for exactly 20 minutes (no more and no less) in a typical week?
   - Write the answer on your sheet of paper that is titled "Exercise 5.2: Answers to Questions."
   - In the frequency table that was generated by PROC FREQ on your SAS output page, circle the number that constitutes your answer. Write "Q4" next to this circled answer.

5. **Question**: How many people indicated that they exercise for 60 minutes or less in a typical week?
   - Write the answer on your sheet of paper that is titled "Exercise 5.2: Answers to Questions."
   - In the frequency table that was generated by PROC FREQ on your SAS output page, circle the number that constitutes your answer. Write "Q5" next to this circled answer.

6. **Question**: What percentage of people exercise for 30 minutes or less in a typical week?
   - Write the answer on your sheet of paper that is titled "Exercise 5.2: Answers to Questions."
   - In the frequency table that was generated by PROC FREQ on your SAS output page, circle the number that constitutes your answer. Write "Q6" next to this circled answer.

7. **Question**: What is the total number of usable observations for the MINUTES variable in this data set?
   - Write the answer on your sheet of paper that is titled "Exercise 5.2: Answers to Questions."
   - In the frequency table that was generated by PROC FREQ on your SAS output page, circle the number that constitutes your answer. Write "Q7" next to this circled answer.

What You Will Hand In

Hand in the following materials stapled together in this order:

1. A printout of your SAS program.
2. A printout of your SAS log.
3. A printout of your SAS output (from PROC FREQ), with the appropriate numbers circled according to the preceding directions.
4. Your sheet of paper titled "Exercise 5.2: Answers to Questions," on which you wrote your answers to the preceding questions.

Hint

If your SAS program ran correctly, your output should resemble the following:

```
                         JANE DOE                                  1

                    The FREQ Procedure

                                           Cumulative    Cumulative
MINUTES    Frequency      Percent         Frequency       Percent
-------------------------------------------------------------------
      0            6        31.58                 6         31.58
     20            4        21.05                10         52.63
     30            1         5.26                11         57.89
     40            1         5.26                12         63.16
     60            2        10.53                14         73.68
    120            1         5.26                15         78.95
    180            2        10.53                17         89.47
    240            1         5.26                18         94.74
    300            1         5.26                19        100.00
```

EXERCISES 6

Exercises for Chapter 6: Creating Graphs

Exercise 6.1: Using PROC CHART to Create Bar Charts from LAT Data

Overview

In this exercise, you will analyze the data set from the fictitious Learning Aptitude Test (LAT) described in the exercises for Chapter 5: Creating Frequency Tables. You will use PROC CHART to create a frequency bar chart for one of the variables in the data set. Within the same program, you will use a second PROC CHART to create a subgroup-mean bar chart from the same data set. Finally, you will write a response on the subgroup-mean bar chart to demonstrate that you understand how to interpret it.

The Study

In this exercise, you will analyze data from the LAT study described in the exercises for Chapter 5. In those exercises, you learned that the LAT is a standardized test that is administered to people who plan to go to college, and its scores are used to make admissions decisions. From a sample study, you obtain scores for 18 subjects based on data from the following four variables:

- **LAT Verbal subtest.** Possible scores can range from 200-800.
- **LAT Math subtest.** Possible scores can range from 200-800.
- **LAT Analytical subtest.** Possible scores can range from 200-800.

- **Area that the subject is majoring in at college.** You use the value "A" to identify subjects majoring in the College of Arts & Sciences, "B" to identify those majoring in the College of Business, and "E" to identify those majoring in the College of Education.

The DATA Step

The DATA step for this exercise should be identical to the DATA step used in Exercise 5.1. If you completed that exercise and saved your SAS program as a computer file, you can open that file, delete the old PROC step (from Exercise 5.1), and append a new PROC step to perform the tasks required in this exercise. As in Exercise 5.1, you will use data from the data set that appeared in Table 5.E1.1. You will also need to specify "LS=80 PS=60" in the OPTIONS line of the PROC step.

If for any reason you do not have the DATA step from Exercise 5.1 available, you will have to write it at this time. To do this, follow the instructions provided in Exercise 5.1 in the sections titled "The Study," "Data Set to be Analyzed," and step 1 of "Your Assignment Regarding the SAS Program."

Your Assignment Regarding the PROC Step of the SAS Program

1. After the DATA step in your program, do the following:
   - Add statements that will cause PROC CHART to create a frequency bar chart for the variable LAT_V (LAT Verbal subtest). Use the DISCRETE option when you create this bar chart. If you need more information, see the relevant section from Chapter 6, "Creating Graphs" of the *Student Guide*.
   - Add a TITLE1 statement after the PROC statement, and type your full name in this TITLE1 statement so that it will appear in the output.
   - As part of the same program, use a separate PROC CHART statement to create a subgroup-mean bar chart. In this chart, assign MAJOR as the "group variable." This will cause the chart to have one bar for students majoring in the College of Arts & Sciences, one bar for

students majoring in the College of Business, and one bar for students majoring in the College of Education.

In this chart, assign LAT_V (LAT Verbal subtest scores) as the "criterion variable." This will cause your bar chart to have "LAT_V Mean" as the label for the vertical axis. The chart will indicate the mean score on the LAT_V for Arts & Sciences majors versus Business majors versus Education majors.

If you need more information, see "Using PROC CHART to Create a Frequency Bar Chart" of the *Student Guide.*

Hint: You will have to make proper use of the VBAR statement with the SUMVAR and TYPE options.

2. Submit the program, and, if necessary, correct it so that it runs without errors.
3. Review your log window and output files for possible errors.

Your Assignment Regarding the SAS Output

After you are sure that the program ran without errors, print your SAS program, SAS log, and SAS output from the two PROC CHART procedures.

The second chart should be the subgroup-means bar chart that plots mean scores on LAT_V for the three types of majors. On this output page, draw an arrow to the bar for the group that has the highest mean score on the LAT_V. Label this arrow "Majors with the highest mean score." Then write the name for this group—either "Arts & Science Majors," "Business Majors," or "Education Majors."

What You Will Hand In

Hand in the following materials stapled in the following order:

1. A printout of your SAS program.
2. A printout of your SAS log.

3. A printout of your SAS output (from both PROC CHART procedures) with the appropriate comments added, according to the preceding directions.

Hint

If your SAS program ran correctly, your output should resemble the following output:

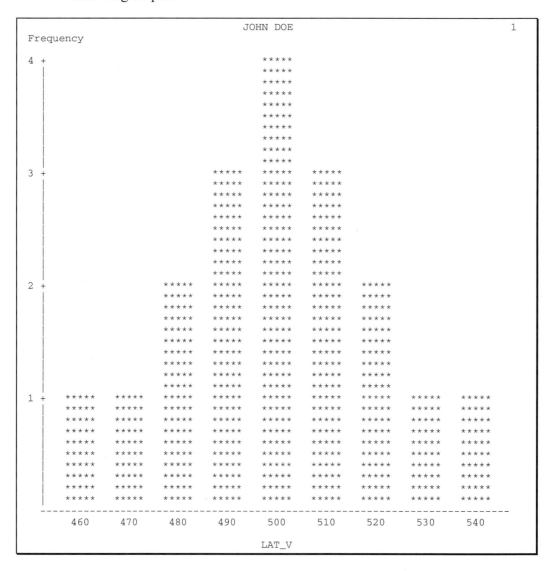

```
                                    JOHN DOE                                       1
Frequency

  4 +                                           *****
    |                                           *****
    |                                           *****
    |                                           *****
    |                                           *****
    |                                           *****
    |                                           *****
    |                                           *****
    |                                           *****
    |                                           *****
  3 +                                 *****     *****     *****
    |                                 *****     *****     *****
    |                                 *****     *****     *****
    |                                 *****     *****     *****
    |                                 *****     *****     *****
    |                                 *****     *****     *****
    |                                 *****     *****     *****
    |                                 *****     *****     *****
    |                                 *****     *****     *****
    |                                 *****     *****     *****
  2 +                       *****     *****     *****     *****     *****
    |                       *****     *****     *****     *****     *****
    |                       *****     *****     *****     *****     *****
    |                       *****     *****     *****     *****     *****
    |                       *****     *****     *****     *****     *****
    |                       *****     *****     *****     *****     *****
    |                       *****     *****     *****     *****     *****
    |                       *****     *****     *****     *****     *****
    |                       *****     *****     *****     *****     *****
    |                       *****     *****     *****     *****     *****
  1 +   *****     *****     *****     *****     *****     *****     *****     *****     *****
    |   *****     *****     *****     *****     *****     *****     *****     *****     *****
    |   *****     *****     *****     *****     *****     *****     *****     *****     *****
    |   *****     *****     *****     *****     *****     *****     *****     *****     *****
    |   *****     *****     *****     *****     *****     *****     *****     *****     *****
    |   *****     *****     *****     *****     *****     *****     *****     *****     *****
    |   *****     *****     *****     *****     *****     *****     *****     *****     *****
    |   *****     *****     *****     *****     *****     *****     *****     *****     *****
    |   *****     *****     *****     *****     *****     *****     *****     *****     *****
    |   *****     *****     *****     *****     *****     *****     *****     *****     *****
    ----------------------------------------------------------------------------------------------
         460       470       480       490       500       510       520       530       540

                                               LAT_V
```

```
                  JOHN DOE                                      2

LAT_V Mean

    |         *****
500 +         *****          *****
    |         *****          *****          *****
    |         *****          *****          *****
    |         *****          *****          *****
    |         *****          *****          *****
450 +         *****          *****          *****
    |         *****          *****          *****
    |         *****          *****          *****
    |         *****          *****          *****
    |         *****          *****          *****
400 +         *****          *****          *****
    |         *****          *****          *****
    |         *****          *****          *****
    |         *****          *****          *****
    |         *****          *****          *****
350 +         *****          *****          *****
    |         *****          *****          *****
    |         *****          *****          *****
    |         *****          *****          *****
    |         *****          *****          *****
300 +         *****          *****          *****
    |         *****          *****          *****
    |         *****          *****          *****
    |         *****          *****          *****
    |         *****          *****          *****
250 +         *****          *****          *****
    |         *****          *****          *****
    |         *****          *****          *****
    |         *****          *****          *****
    |         *****          *****          *****
200 +         *****          *****          *****
    |         *****          *****          *****
    |         *****          *****          *****
    |         *****          *****          *****
    |         *****          *****          *****
150 +         *****          *****          *****
    |         *****          *****          *****
    |         *****          *****          *****
    |         *****          *****          *****
    |         *****          *****          *****
100 +         *****          *****          *****
    |         *****          *****          *****
    |         *****          *****          *****
    |         *****          *****          *****
    |         *****          *****          *****
 50 +         *****          *****          *****
    |         *****          *****          *****
    |         *****          *****          *****
    |         *****          *****          *****
    |         *****          *****          *****
    -------------------------------------------------------
                A              B              E

                             MAJOR
```

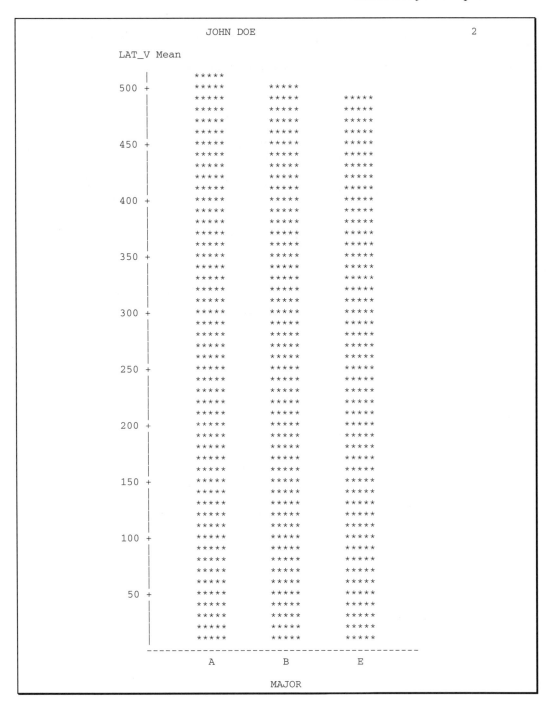

Exercise 6.2: Using PROC CHART to Create Bar Charts from an Experiment About Goal Setting

Overview

In this exercise, you will analyze data from a fictitious experiment about goal setting. You will use PROC CHART to create a frequency bar chart to determine how many subjects are in each treatment condition. You will also use PROC CHART to create a subgroup-mean bar chart to determine which experimental condition group scored highest on the study's dependent variable. Finally, you will write responses on the bar charts to demonstrate that you understand how to interpret them.

The Study

Background. Suppose that you are conducting research about organizational behavior. You are particularly interested in finding interventions that are effective in increasing safety in the work place. In your current project, you are working with pizza deliverers who work for a large home-delivery pizza franchise. You want to compare the effectiveness of different approaches for promoting safe driving behaviors on the part of these deliverers.

Goal setting theory (e.g., Lock & Latham, 1990) predicts that task performance tends to be better when workers are given challenging work-related goals. In your current experiment, you will determine whether pizza deliverers will display an improvement in safe driving behaviors when given specific goals related to their driving. The driving behavior that you are focusing on in the current study involves *stopping at stop signs.* You want to increase the likelihood that the drivers will come to a complete stop at stop signs (as opposed to coming to a "rolling stop"). Your study is designed to determine whether goal setting can help achieve the desired result.

Manipulating the independent variable. In this study, you begin with a group of 27 pizza deliverers (subjects), and you randomly assign each subject to one of three condition groups:

- **Participative goal setting condition group**. Nine subjects are assigned to the participative goal setting condition group. You meet with these drivers and review data that shows that they typically come to complete stops about 50 percent of the time. You ask them to decide upon a reasonable goal for improving their performance. In the discussion, the group decides on a goal of 75 percent. They agree to attempt to come to a complete stop at stop signs at least 75 percent of the time.
- **Assigned goals condition group**. Eight subjects are assigned to the assigned goals condition group. You meet with these drivers, and again review data that shows that they come to complete stops about 50 percent of the time. You assign a goal of 75 percent to this group. You ask them to attempt to come to a complete stop at stop signs at least 75 percent of the time. Notice that the goal for this group is identical to the goal for the participative condition group. The difference is that, in the participative condition group, subjects set the goal for themselves; whereas in the assigned condition group, the goal is assigned to the subjects without participation.
- **Control condition group**. Ten subjects are assigned to the control condition group. No driving-related goals are set for subjects in this group.

Measuring the dependent variable. The dependent variable in this study is the number of complete stops demonstrated by each subject out of 30 opportunities. You position observers at a stop sign near the pizza franchise and instruct them to unobtrusively observe the pizza deliverers from your study when they come to the stop sign. The observers (a) record whether the deliverers come to a complete stop at the sign, and (b) also record the driver's license tag number. Later, you use the tag numbers to determine the assigned treatment condition group for each driver.

For each driver, you record how many times he or she came to a complete stop at the stop sign out of 30 opportunities. Therefore, each driver's score on this dependent variable could range from zero (if the driver never came to a complete stop) to 30 (if the driver always came to a complete stop).

Note: Although the study and results described here are fictitious, they are inspired by the actual study reported by Ludwig and Geller (1997).

Data Set to be Analyzed

Table 6.E2.1 provides the data set that you will analyze in this exercise.

Each horizontal row in Table 6.E2.1 presents data for a different subject. For example, the first row presents data for Subject number "01" who has a value of "P" for the Condition variable and a value of "27" for the Number of stops variable.

The Subject number column in Table 6.E2.1 lists a unique number for each subject.

The Condition column indicates the experimental condition group to which a subject was assigned. The letter "P" identifies subjects in the participative goal setting condition group, the letter "A" identifies subjects in the assigned goal setting condition group, and the letter "C" identifies subjects in the control condition group. You can see that subjects 1-9 were in the participative goal setting condition group, subjects 10-17 were in the assigned goal setting condition group, and subjects 18-27 were in the control condition group.

The Number of stops column indicates each subject's score on the study's dependent variable—the number of complete stops they displayed out of 30 opportunities. You can see that subject number "01"came to 27 complete stops, subject number "02" came to 25 complete stops, and so forth.

Table 6.E2.1

Data from the Goal Setting Experiment

Subject number	Condition	Number of stops
01	P	27
02	P	25
03	P	23
04	P	20
05	P	25
06	P	30
07	P	27
08	P	19
09	P	25
10	A	30
11	A	23
12	A	22
13	A	20
14	A	22
15	A	17
16	A	25
17	A	26
18	C	22
19	C	17
20	C	17
21	C	13
22	C	10
23	C	16
24	C	15
25	C	13
26	C	18
27	C	20

Your Assignment Regarding the SAS Program

1. Write a SAS program that will input and analyze the preceding data set. When you write this program, do the following:
   - Use the DATA step from the SAS program that was presented in Chapter 5, "Creating Frequency Tables" of the *Student Guide* as a

model. If necessary, you can also refer to Chapter 4, "Data Input" for more information.

- So that your output will resemble the output presented in the following "Hint" section, use the following OPTIONS statement:

```
OPTIONS   LS=80   PS=60;
```

  However, if your charts are "broken" across two output pages (when printed), use the following OPTIONS statement to change the page size that you request:

```
OPTIONS   LS=80   PS=50;
```

  Note: If you use the second of these two OPTIONS statements, your output might not be identical to the output presented in the following "Hint" section.

- Type all of the information from Table 6.E2.1. You should type the Subject number variable as the first column as it appears in Table 6.E2.1. When you type subject numbers such as 01, make sure that you type a zero (0) and not the letter O.
- Assign the following SAS variable names to the variables.
  - Use the SAS variable name SUB_NUM to represent Subject numbers (the first column in Table 6.E2.1). For this variable name, make sure that you use an underscore (_), and not a hyphen (-).
  - Use the SAS variable name COND for the variable that represents the experimental conditions to which the subjects were assigned.
  - Use the SAS variable name STOPS for the variable that indicates how many complete stops each subject made.

2. After you write the DATA step, do the following:

- Add statements that will cause PROC CHART to create a frequency bar chart for the variable COND. If you do this correctly, the resulting bar chart will indicate how many subjects were in the participative goal setting condition group, how many were in the assigned goals condition group, and how many were in the control condition group. If you need more information, see the relevant section from Chapter 6, "Creating Graphs" of the *Student Guide*.

- Add a TITLE1 statement after this PROC statement, and type your full name in the TITLE1 statement so that it will appear in the output.
- As part of the same program, use a separate PROC CHART statement to create a subgroup-mean bar chart.
  - The "group variable" should be COND so that your chart will have one bar for subjects in the participative goal setting group, one bar for subjects in the assigned goals group, and one bar for subjects in the control group.
  - The "criterion variable" should be STOPS (number of complete stops made by subjects) so that your bar chart will have "STOPS Mean" as the label for the vertical axis. The chart will indicate the mean number of stops made by subjects in the participative goal setting group versus the assigned goals group versus the control group.
  - If you need more information, see "Using PROC CHART to Create a Frequency Bar Chart" from the *Student Guide.*
  - Hint: You will have to make proper use of the VBAR statement using the SUMVAR and TYPE options.

3. Submit the program, and, if necessary, correct it so that it runs without errors.
4. Review your log window and output files for possible errors.

Your Assignment Regarding the SAS Output

After you are sure that the program ran correctly, print your SAS program, SAS log, and SAS output from the two PROC CHART procedures. Then do the following:

1. The first chart included in your SAS output should be the frequency bar chart that indicates the number of people in each of the three groups. On this output page, draw an arrow to the bar for the group that has the largest frequency. Label this arrow "Group with the most subjects." Then write the name for this group—either "Participative Goal Setting Condition Group," "Assigned Goals Condition Group," or "Control Condition Group."

2. The second chart included in your SAS output should be the subgroup-means bar chart that plots mean scores on STOPS for the three groups. On this output page, draw an arrow to the bar for the group that has the highest mean score on the STOPS dependent variable. Label this arrow "Group with the highest mean score." Then write the name for this group—either "Participative Goal Setting Group," "Assigned Goals Group," or "Control Group."

What You Will Hand In

Hand in the following materials stapled together in the following order:

1. A printout of your SAS program.
2. A printout of your SAS log.
3. A printout of your SAS output (from both PROC CHART procedures) with the appropriate comments added, according to the preceding directions.

Hint

If your SAS program ran correctly, your output should resemble the following:

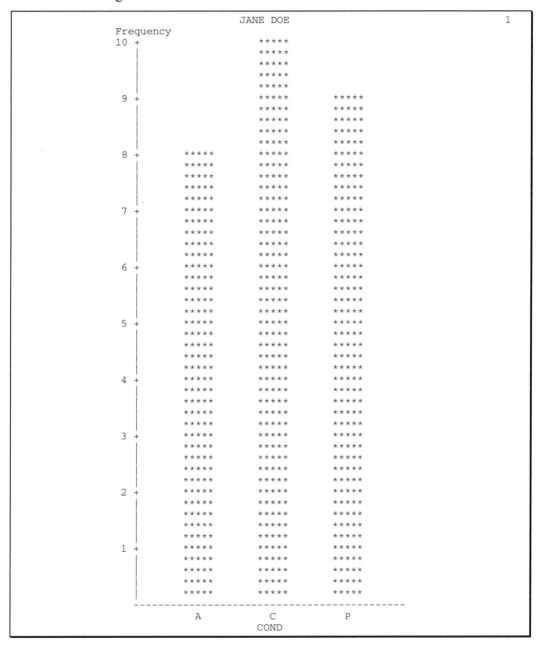

```
                    JANE DOE                                   1
Frequency
10 +                   *****
   |                   *****
   |                   *****
   |                   *****
   |                   *****
 9 +                   *****       *****
   |                   *****       *****
   |                   *****       *****
   |                   *****       *****
   |                   *****       *****
 8 +       *****       *****       *****
   |       *****       *****       *****
   |       *****       *****       *****
   |       *****       *****       *****
   |       *****       *****       *****
 7 +       *****       *****       *****
   |       *****       *****       *****
   |       *****       *****       *****
   |       *****       *****       *****
   |       *****       *****       *****
 6 +       *****       *****       *****
   |       *****       *****       *****
   |       *****       *****       *****
   |       *****       *****       *****
   |       *****       *****       *****
 5 +       *****       *****       *****
   |       *****       *****       *****
   |       *****       *****       *****
   |       *****       *****       *****
   |       *****       *****       *****
 4 +       *****       *****       *****
   |       *****       *****       *****
   |       *****       *****       *****
   |       *****       *****       *****
   |       *****       *****       *****
 3 +       *****       *****       *****
   |       *****       *****       *****
   |       *****       *****       *****
   |       *****       *****       *****
   |       *****       *****       *****
 2 +       *****       *****       *****
   |       *****       *****       *****
   |       *****       *****       *****
   |       *****       *****       *****
   |       *****       *****       *****
 1 +       *****       *****       *****
   |       *****       *****       *****
   |       *****       *****       *****
   |       *****       *****       *****
   |       *****       *****       *****
   --------------------------------------------
             A           C           P
                       COND
```

```
                              JANE DOE                                   2
 STOPS Mean
    |                                        *****
 24 +                                        *****
    |                                        *****
    |      *****                             *****
    |      *****                             *****
 22 +      *****                             *****
    |      *****                             *****
    |      *****                             *****
    |      *****                             *****
 20 +      *****                             *****
    |      *****                             *****
    |      *****                             *****
    |      *****                             *****
 18 +      *****                             *****
    |      *****                             *****
    |      *****                             *****
    |      *****                             *****
 16 +      *****        *****                *****
    |      *****        *****                *****
    |      *****        *****                *****
    |      *****        *****                *****
 14 +      *****        *****                *****
    |      *****        *****                *****
    |      *****        *****                *****
    |      *****        *****                *****
 12 +      *****        *****                *****
    |      *****        *****                *****
    |      *****        *****                *****
    |      *****        *****                *****
 10 +      *****        *****                *****
    |      *****        *****                *****
    |      *****        *****                *****
    |      *****        *****                *****
  8 +      *****        *****                *****
    |      *****        *****                *****
    |      *****        *****                *****
    |      *****        *****                *****
  6 +      *****        *****                *****
    |      *****        *****                *****
    |      *****        *****                *****
    |      *****        *****                *****
  4 +      *****        *****                *****
    |      *****        *****                *****
    |      *****        *****                *****
    |      *****        *****                *****
  2 +      *****        *****                *****
    |      *****        *****                *****
    |      *****        *****                *****
    |      *****        *****                *****
 --------------------------------------------------------------
             A            C                    P
                         COND
```

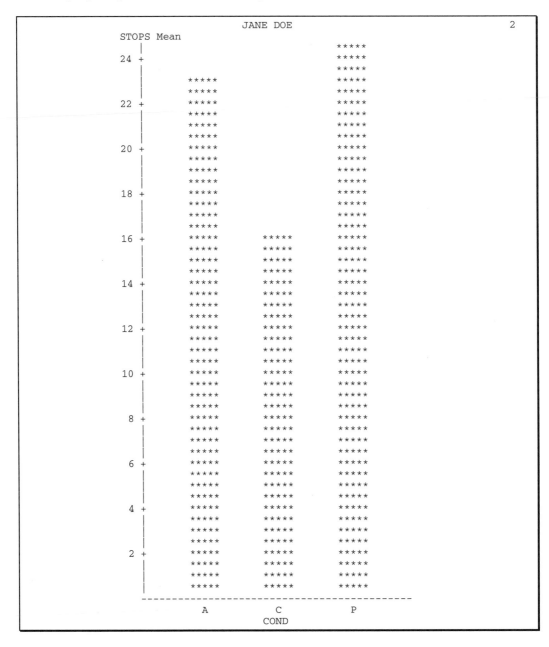

EXERCISES

Exercises for Chapter 7: Measures of Central Tendency and Variability

7

Exercise 7.1: Using PROC UNIVARIATE to Identify Normal, Skewed, and Bimodal Distributions

Overview

In this exercise, you will analyze the data set from the fictitious Learning Aptitude Test (LAT) described in Chapter 5, "Creating Frequency Tables." You will use PROC UNIVARIATE to compute descriptive statistics and create stem-and-leaf plots for three variables in the data set. You will prepare a table that summarizes some measures of central tendency and variability, and will also identify the shape of the distributions presented in the stem-and-leaf plots.

The Study

In this exercise, you will analyze data from the LAT study described in Exercise 5.1 from Chapter 5, "Creating Frequency Tables." There, you learned that the LAT is a fictitious standardized test that is administered to people who plan to go to college, and its scores are used to make admissions decisions.

From a sample study of 18 college students (subjects), you obtain scores for 18 subjects based on data from the following four variables:

- **LAT Verbal subtest.** Possible scores can range from 200-800. You will use the SAS variable name LAT_V to represent this variable.
- **LAT Math subtest.** Possible scores can range from 200-800. You will use the SAS variable name LAT_M to represent this variable.
- **LAT Analytical subtest.** Possible scores can range from 200-800. You will use the SAS variable name LAT_A to represent this variable.
- **Area that the subject is majoring in at college.** You will use the value "A" to identify students majoring in the College of Arts & Sciences, "B" to identify those majoring in the College of Business, and "E" to identify those majoring in the College of Education. You will use the SAS variable name MAJOR to represent this variable.

The DATA Step

The DATA step for this exercise should be identical to the DATA step used with Exercise 5.1 from the Chapter 5 exercises. If you completed that exercise and saved your SAS program as a computer file, you should open that file, delete the old PROC step (from Exercise 5.1), and append a new PROC step to perform the tasks required in this exercise. Once again, you will analyze the data set that appeared in Table 5.E1.1 in the Chapter 5 exercises.

If for any reason you do not have the DATA step from Exercise 5.1 available, you will have to prepare it at this time. To do this, follow the instructions provided in Exercise 5.1 in the sections headed "The Study," "Data Set to be Analyzed," and step 1 only (the step that discusses the DATA step) of "Your Assignment Regarding the SAS Program."

Your Assignment Regarding the PROC Step of the SAS Program

1. After the DATA step, do the following:
   - Add statements that perform PROC UNIVARIATE on the variables LAT_V, LAT_M, and LAT_A.
   - Request the PLOT and NORMAL options. If you need more information, see the relevant section from Chapter 7, "Measures of Central Tendency and Variability" in the *Student Guide*.
2. Submit the program, and, if necessary, correct it so that it runs without errors.
3. Review your log window and output files for possible errors. Print your SAS program, SAS log, and SAS output.

Your Assignment Regarding the Output

You should read all of the following directions before you begin the assignment.

Review your SAS output. Verify that you have one set of PROC UNIVARIATE output for LAT_V, one set for LAT_M, and one set for LAT_A. For each variable analyzed, you should have approximately two pages of output. The "Hint" section near the end of this exercise shows the type of information that you should have for each variable.

Item 1

The purpose of this part of the assignment is to verify that you can find the mean, median, mode, standard deviation, and sample size in your output.

Take out a blank sheet of paper. At the top, write (a) your full name, and (b) the title Exercise 7.1. Then, below that, write the heading "Exercise 7.1, Item 1," and copy the following table onto your sheet of paper (be sure to write small and leave plenty of space—you are going to copy numbers below each of the statistical headings in your table).

Table 7.E1.1

Statistics for the LAT Study

Variable	M	Mdn	Mode	SD	N
LAT Verbal					
LAT Math					
LAT Analytical					

The six headings that appear in Table 7.E1.1 are described as follows:

- The first heading in the table is "Variable." Below this heading are the names of the three variables that you analyzed using PROC UNIVARIATE: LAT Verbal, LAT Math, and LAT Analytical.
- The second heading is "M." This is the statistical symbol for the sample mean.
- The third heading is "Mdn." This is the abbreviation for the median.
- The fourth heading is Mode, the most frequently occurring score. Notice that this heading is not underlined because it is neither a symbol nor an abbreviation.
- The fifth heading is "SD." This is the abbreviation for standard deviation.
- The sixth heading is "N." This is the symbol for the total number of subjects in a sample.

After you have finished copying your table, review your SAS output that provides results from the PROC UNIVARIATE analysis of LAT_V (the SAS variable name for the LAT Verbal subtest). Find the mean score on LAT_V. In your table, write this mean at the location where the row named LAT Verbal intersects with the column named M. Now repeat this process for the remaining statistics in Table 7.E1.1 (in other words, find the median for LAT Verbal, and write it under the heading Mdn, and then find the mode for LAT Verbal, and write it under the heading Mode, and so forth).

When filling in the entries in your table, if any statistics have decimal places, you should round up to two decimal places.

After you have finished filling in all of the statistics for LAT_V, review your SAS output that provides results from the PROC UNIVARIATE analysis of LAT_M (the LAT Math subtest). Find the mean score on LAT_M. In your table, write this mean at the location where the row named LAT Math intersects with the column named M. Now repeat this process for all of the remaining statistics in the row named LAT Math.

Finally, review your SAS output that provides the results from the PROC UNIVARIATE analysis of LAT_A (the LAT Analytical subtest). Find the mean score on LAT_A. In your table, write this mean at the location where the row named LAT Analytical intersects with the column named M. Repeat this process for all of the remaining statistics in the row named LAT Analytical.

When you are finished, Table 7.E1.1 should be completely filled in. That is, it should contain all five statistics for all three variables.

Item 2

The purpose of this part of the exercise is to verify that you can review a stem-and-leaf plot and correctly label the general shape of a distribution.

A) Review the stem-and-leaf plot for the variable LAT_V. Determine the general shape of this distribution. Then, on the same sheet of paper you used for Item 1, write the heading: "Item 2.A." To the right of this heading, select and write a statement from the following list that best describes the LAT_V plot:

- "The distribution for LAT Verbal is approximately normal"
- "The distribution for LAT Verbal is positively skewed"
- "The distribution for LAT Verbal is negatively skewed"
- "The distribution for LAT Verbal is bimodal"

Below your statement, write a brief explanation (1-3 sentences) explaining why you believe that this statement is correct. For example, if you said that "The distribution for LAT Verbal is positively skewed," explain *why* you believe that it is positively skewed. What is it about the shape of this distribution that tells you that it is positively skewed? Chapter 7 of the *Student Guide* provides definitions for some of the relevant terms (such as "positive skewness," "negative skewness," and "bimodal distribution"). Your explanations should include these definitions, when appropriate.

B) Review the stem-and-leaf plot for the variable LAT_M. Determine the general shape of this distribution. Then, on the same sheet of paper, write "Item 2.B" below your answer for Item 2.A. To the right of this heading, select and write a statement from the following list that best describes the LAT_M plot:

- "The distribution for LAT Math is approximately normal"
- "The distribution for LAT Math is positively skewed"
- "The distribution for LAT Math is negatively skewed"
- "The distribution for LAT Math is bimodal"

Below your statement, write a brief explanation (1-3 sentences) explaining why you believe that this statement is correct.

C) Review the stem-and-leaf plot for the variable LAT_A. Determine the general shape of this distribution. Then, on the same sheet of paper, write the heading "Item 2.C" below your answer for Item 2.B. To the right of this heading, select and write a statement from the following list that best describes the LAT_A plot:

- "The distribution for LAT Analytical is approximately normal"
- "The distribution for LAT Analytical is positively skewed"
- "The distribution for LAT Analytical is negatively skewed"
- "The distribution for LAT Analytical is bimodal"

Below your statement, write a brief explanation (1-3 sentences) explaining why you believe that this statement is correct.

If you need information regarding these tasks, see the section "Interpreting a Stem-and-Leaf Plot Created by PROC UNIVARIATE" from Chapter 7 of the *Student Guide*.

What You Will Hand In

Hand in the following materials stapled together in this order:

1. A printout of your SAS program.
2. A printout of your SAS log.
3. A printout of your SAS output.
4. A sheet of paper with your table from Item 1 and your responses to Item 2.

Hint

If your SAS program ran correctly, your first set of output (for the variable LAT_V) should resemble the following output. Remember that your program will produce more output than shown in the excerpt reproduced below.

```
                              JOHN DOE                              1

                      The UNIVARIATE Procedure
                          Variable:  LAT_V

                              Moments

N                          18    Sum Weights                 18
Mean                      500    Sum Observations          9000
Std Deviation      20.5798302    Variance            423.529412
Skewness                    0    Kurtosis            -0.1357639
Uncorrected SS        4507200    Corrected SS              7200
Coeff Variation    4.11596604    Std Error Mean       4.8507125


                     Basic Statistical Measures

         Location                    Variability

    Mean     500.0000     Std Deviation            20.57983
    Median   500.0000     Variance                423.52941
    Mode     500.0000     Range                    80.00000
                          Interquartile Range      20.00000
```

Continued on next page

Continued from previous page

```
                 Tests for Location: Mu0=0

     Test           -Statistic-    -----p Value------

     Student's t    t  103.0776    Pr > |t|     <.0001
     Sign           M         9    Pr >= |M|    <.0001
     Signed Rank    S      85.5    Pr >= |S|    <.0001


                    Tests for Normality

Test                  --Statistic---    -----p Value------

Shapiro-Wilk          W     0.983895    Pr < W      0.9812
Kolmogorov-Smirnov    D     0.111111    Pr > D     >0.1500
Cramer-von Mises      W-Sq  0.036122    Pr > W-Sq  >0.2500
Anderson-Darling      A-Sq  0.196144    Pr > A-Sq  >0.2500


                    Quantiles (Definition 5)
                   Quantile          Estimate

                   100% Max             540
                    99%                 540
                    95%                 540
                    90%                 530
                    75% Q3              510
                    50% Median          500
                    25% Q1              490
                    10%                 470
                     5%                 460
                     1%                 460
                     0% Min             460
```

```
                    JOHN DOE                              2

              The UNIVARIATE Procedure
                  Variable:  LAT_V

                Extreme Observations

        ----Lowest----          ----Highest---

        Value      Obs          Value      Obs

          460       18            510       11
          470       17            520        5
          480       16            520        6
          480       14            530        8
          490       15            540        1


     Stem Leaf                   #            Boxplot
       54 0                      1               |
       53 0                      1               |
       52 00                     2               |
       51 000                    3            +-----+
       50 0000                   4            *--+--*
       49 000                    3            +-----+
       48 00                     2               |
       47 0                      1               |
       46 0                      1               |
          ----+----+----+----+
      Multiply Stem.Leaf by 10**+1


                  Normal Probability Plot
     545+                                             * +++++
        |                                        * +++++
        |                                   * +*+++
        |                               **+*++
     505+                         **+**++
        |                    *  **+++
        |                *  +*+++
        |             +*+++
     465+        +*+++
         +----+----+----+----+----+----+----+----+----+----+
             -2        -1         0        +1        +2
```

Exercise 7.2: Using PROC UNIVARIATE and PROC MEANS to Compute Measures of Variability

Overview

In this exercise, you will analyze the data set from the fictitious Learning Aptitude Test (LAT) described in Chapter 5, "Creating Frequency Tables." You will use PROC UNIVARIATE to compute descriptive statistics and create a stem-and-leaf plot for one of the variables in the data set. You will also use PROC MEANS to compute the sample variance and standard deviation, along with the estimated population variance and standard deviation, for one of the variables. You will prepare two tables of statistics to demonstrate that you are able to interpret the SAS output that you have created.

The Study

In this exercise, you will analyze data from the LAT study described in Exercise 5.1 from Chapter 5, "Creating Frequency Tables." There, you learned that the LAT is a fictitious standardized test that is administered to people who plan to go to college, and its scores are used to make admissions decisions.

From a sample of 18 college students (subjects), you obtain scores for 18 subjects based on data from the following four variables:

- **LAT Verbal subtest.** Possible scores can range from 200-800. You will use the SAS variable name LAT_V to represent this variable.
- **LAT Math subtest.** Possible scores can range from 200-800. You will use the SAS variable name LAT_M to represent this variable.
- **LAT Analytical subtest.** Possible scores can range from 200-800. You will use the SAS variable name LAT_A to represent this variable.
- **Area that the subject is majoring in at college.** You will use the value "A" to identify students majoring in the College of Arts & Sciences, "B" to

identify those majoring in the College of Business, and "E" to identify those majoring in the College of Education. You will use the SAS variable name MAJOR to represent this variable.

The DATA Step

The DATA step for this exercise should be identical to the DATA step used with Exercise 5.1 from the Chapter 5 exercises. If you completed that exercise and saved your SAS program as a computer file, you should open that file, delete the old PROC step (from Exercise 5.1), and append a new PROC step to perform the tasks required in this exercise. Once again, you will analyze the data set that appeared in Table 5.E1.1 in the Chapter 5 exercises.

If for any reason you do not have the DATA step from Exercise 5.1 available, you will have to prepare it at this time. To do this, following the instructions provided in Exercise 5.1 in the sections headed "The Study," "Data Set to be Analyzed," and step 1 only (the step that discusses the DATA step) of "Your Assignment Regarding the Program."

Your Assignment Regarding the PROC Step of the SAS Program

1. After the DATA step, add some statements that will do all of the following in a single program:
   - Cause PROC UNIVARIATE to be performed on LAT_V.
   - Cause PROC MEANS to be performed on LAT_V so that it computes the sample standard deviation and sample variance (along with the mean and other statistics illustrated in the section titled "Computing the Sample Variance and Standard Deviation" from Chapter 7, "Measures of Central Tendency and Variability" in the *Student Guide*). Remember that the sample standard deviation and variance are the ones that use N as the divisor.
   - In the TITLE1 statement following the preceding PROC MEANS statement, include your full name along with the words "DIVISOR =

N". This will help you remember that this output provides the sample standard deviation.

- Cause PROC MEANS to be performed on LAT_V a second time so that it computes the estimated population standard deviation and estimated population variance (along with the mean and other statistics illustrated in the section titled "Computing the Estimated Population Variance and Standard Deviation" in the *Student Guide*). Remember that the estimated population standard deviation and variance are the ones that use N – 1 as the divisor.
- In the TITLE1 statement following the preceding PROC MEANS statement, include your full name along with the words "DIVISOR = N – 1". This will help you remember that this output provides the estimated population standard deviation.

2. Submit the program, and, if necessary, correct it so that it runs without errors.
3. Review your log window and output files for possible errors. Print your SAS program, SAS log, and SAS output.

Your Assignment Regarding the Output

You should read all of the following directions before you start writing.

Review your SAS output. Verify that you have PROC UNIVARIATE output for LAT_V. In addition, you should have two sets of PROC MEANS output for LAT_V. The "Hint" section near the end of this exercise shows the type of information that you should have for each variable.

Item 1

The purpose of this part of the assignment is to verify that you can find various measures of variability in your PROC UNIVARIATE output.

Take out a blank sheet of paper. At the top, write (a) your full name, and (b) the title Exercise 7.2. Then, below that, write the heading "Exercise 7.2, Item 1," and copy the following table onto your sheet of paper (be sure to write small and leave plenty of space—you are going to copy numbers below each of the statistical headings in your table).

Table 7.E2.1

Simple Measures of Variability for the LAT Verbal Subtest

Variable	Range	Q_1	Q_3	Interquartile Range	Semi-interquartile Range
LAT Verbal					

The six headings that appear in Table 7.E2.1 are described as follows:

- The first heading in the table is "Variable." Below this heading is the name of the variable that you analyzed using PROC UNIVARIATE: LAT Verbal.
- The second heading is "Range," which is self-explanatory.
- The third heading is "Q_1," which is the symbol for the first quartile.
- The fourth heading is "Q_3," which is the symbol for the third quartile.
- The fifth heading is "Interquartile Range," which is self-explanatory.
- The last heading is "Semi-interquartile range," which should also be self-explanatory.

After you have finished copying your table, review your SAS output that provides results from the PROC UNIVARIATE analysis of LAT_V (the SAS variable name for the LAT Verbal subtest). On your PROC UNIVARIATE output, find the various measures of variability that appear as headings in Table 7.E2.1. Copy those statistics onto the appropriate locations in the blank table that you have just created. As you do this, keep the following in mind:

- For more information, refer to the section titled "Simple Measures of Variability: The Range, the Interquartile Range, and the Semi-Interquartile Range" in Chapter 7 of the *Student Guide*.
- The last statistic in the table—the semi-interquartile range—does not actually appear in your SAS output. You will have to compute this by hand.

For more information, see the section titled "The Semi-Interquartile Range" in Chapter 7 of the *Student Guide*.

- When you are finished, Table 7.E2.1 should be completely filled in.

Item 2

The purpose of this part of the exercise is to verify that you can correctly interpret SAS output that provides the sample standard deviation and the estimated population standard deviation. You should read all of the following directions before you start writing.

On the same sheet of paper you used for Item 1, write the heading "Exercise 7.2, Item 2" and copy the following table onto your sheet of paper (be sure to write small and leave plenty of space—you are going to copy numbers below each of the statistical headings in your table).

Table 7.E2.2

More Complex Measures of Variability for the LAT Verbal Subtest

Variable	Sample SD	Estimated population SD
LAT Verbal		

The three headings that appear in Table 7.E2.2 are described as follows:

- The first heading in the table is "Variable." Below this heading is the name of the variable that you analyzed using PROC MEANS: LAT Verbal.
- The second heading is "Sample SD," which stands for sample standard deviation.
- The third heading is "Estimated population SD," which stands for estimated population standard deviation.

After you have finished copying your table, review your SAS output that provides results from the two PROC MEANS analyses of LAT_V. On your PROC MEANS output, find the measures of variability that appear as headings in Table 7.E2.2. Copy those statistics onto the appropriate locations in the blank table that you have just created. For both statistics, round to two decimal places.

In the space below your table, provide an answer to the following question:

> Consider the size of the sample SD versus the size of the estimated population SD that you have just copied into your table. Which of these two statistics is larger? Does it make *sense* that this statistic is larger, given the formulas for the two statistics? In your answer, you must discuss (a) the divisor in the formula for the sample SD, and (b) the divisor in the formula for the estimated population SD.

Be sure to print your answer so that it is easy to read.

What You Will Hand In

Hand in the following materials stapled together in this order:

1. A printout of your SAS program.
2. A printout of your SAS log.
3. A printout of your SAS output.
4. A sheet of paper with your tables and responses.

Hints

If your SAS program ran correctly, your output for the two PROC MEANS should resemble the following output. Remember that, in addition to the following output, you should also have output from PROC UNIVARIATE.

```
                     JOHN DOE... DIVISOR = N                          9

                        The MEANS Procedure

                     Analysis Variable : LAT_V

 N          Mean       Std Dev       Variance        Minimum         Maximum
------------------------------------------------------------------------------
18   500.0000000   20.0000000   400.0000000   460.0000000   540.0000000
------------------------------------------------------------------------------
```

```
                    JOHN DOE... DIVISOR = N-1                        10

                       The MEANS Procedure

                    Analysis Variable : LAT_V

 N          Mean       Std Dev       Variance        Minimum         Maximum
------------------------------------------------------------------------------
18   500.0000000   20.5798302   423.5294118   460.0000000   540.0000000
------------------------------------------------------------------------------
```

EXERCISES 8

Exercises for Chapter 8: Creating and Modifying Variables and Data Sets

Exercise 8.1: Working with an Academic Development Questionnaire

Overview

In this exercise, you will analyze data from a fictitious investigation of academic development in college students. You will write a SAS program that (a) inputs a data set, (b) prints the raw data, (c) performs a variety of data manipulations on the data set, and (d) prints out the new, modified data set.

The Study

Suppose that you are conducting research dealing with academic development in college students. You have developed a self-report measure of academic development that can be administered to college students. The questionnaire follows:

Directions: Please indicate the extent to which you agree or disagree with each of the following statements. You will do this by circling the appropriate number to the left of that statement. The following format shows what each response number stands for:

7 = Agree Very Strongly
6 = Agree Strongly
5 = Agree Somewhat
4 = Neither Agree nor Disagree
3 = Disagree Somewhat
2 = Disagree Strongly
1 = Disagree Very Strongly

For example, if you "Disagree Very Strongly" with the first statement, circle the "1" to the left of that statement. If you "Agree Somewhat," circle the "5," and so forth.

Circle Your Response	
1 2 3 4 5 6 7	1. I am satisfied with the extent of my academic development since enrolling at this university.
1 2 3 4 5 6 7	2. I am satisfied with my academic experience at this university.
1 2 3 4 5 6 7	3. My interest in intellectual matters has decreased since I enrolled at this university.
1 2 3 4 5 6 7	4. Most of my courses at this university have been intellectually stimulating.

5. What is your sex? (circle one)   Female (F)   Male (M)

6. What is your age in years? ________________

7. What is your classification? (circle one)

Freshman (1)   Sophomore (2)   Junior (3)   Senior (4)

The most important items on this questionnaire are items 1-4; these are the statements designed to assess academic development. Items 1, 2, and 4 were written so that, the more a student agrees with them, the greater that student's level of academic development. Item 3, in contrast, is a reversed item—the more a student agrees with item 3, the lower that student's level of academic development.

Data Set to be Analyzed

Suppose that you administer this questionnaire to a sample of 12 college students (subjects). Their responses appear in Table 8.E1.1 as follows:

Table 8.E1.1

Data from the Academic Development Study

	Agree-disagree questions						
Subject number	Q1	Q2	Q3	Q4	Sex	Age	Class
01	6	.	2	7	F	20	1
02	3	2	7	2	M	26	1
03	7	7	1	7	M	19	1
04	5	6	.	5	F	23	2
05	6	7	1	6	M	21	2
06	3	2	6	3	F	25	2
07	5	6	2	5	F	25	3
08	5	6	1	5	F	23	3
09	7	7	1	6	M	31	3
10	5	4	1	4	M	25	4
11	4	5	3	5	F	42	4
12	7	6	1	6	F	38	4

The contents of Table 8.E1.1 should be self-explanatory. In the Subject Number column, a unique subject number is assigned to each of the 12 subjects. The Q1 column contains subject responses to item 1 from the questionnaire, the Q2 column contains responses to item 2, and so forth.

The last column (Class) contains responses to item 7 from the questionnaire, indicating each subject's student classification. The value "1" identifies freshmen, the value "2" identifies sophomores, the value "3" identifies juniors, and the value "4" identifies seniors.

Your Assignment

In this exercise, you will write a single SAS program that performs a number of data manipulation tasks. This program will also include PROC PRINT statements that will print the raw data from the data sets that you create. The most important thing to remember is that all of the following tasks must be performed within a single SAS program.

Use the SAS programs presented in Chapter 8 of the *Student Guide* as models. A program that you should find particularly useful appears in the section titled "Combining a Large Number of Data Manipulation and Data Subsetting Statements In a Single Program."

In the SAS program that you write, do the following:

1. Create a SAS program that will input the data set presented in Table 8.E1.1, which you will title "D1." When writing the initial DATA step do the following:
   - When typing your data, type all of the information from Table 8.E1.1. You should type the Subject number variable that appears as the first column in Table 8.E1.1. When typing subject numbers such as 01, make sure that you type a zero (0) and not the letter O.
   - Use the following SAS variable names in the INPUT statement: SUB_NUM, Q1, Q2, Q3, Q4, SEX, AGE, and CLASS.
2. Following the initial DATA step, use PROC PRINT to print out the raw data (all variables). Include your full name in the TITLE1 statement.
3. As part of the same program, create a new data set titled "D2," and make it a duplicate of D1. In this new DATA step, do the following:
   - Recode variable Q3 so that it is no longer a reversed variable.

- Create a new variable named DEVEL. A given subject's score for DEVEL should be the average of his or her responses on Q1, Q2, Q3 and Q4.
- Create a new numeric variable named AGE2. If a subject's value for AGE is less than 30, then that subject's value for AGE2 should be zero (0). If a subject's value for AGE is greater than or equal to 30, then that subject's value for AGE2 should be 1.
- Create a new character variable named CLASS2. CLASS2 should be created so that its values can be up to 3 characters long. Values for CLASS2 should be assigned in the following way:
  - If a given subject's value for CLASS is 1, then that subject's value for CLASS2 should be FRE (for "freshman").
  - If a given subject's value for CLASS is 2, then that subject's value for CLASS2 should be SOP (for "sophomore").
  - If a given subject's value for CLASS is 3, then that subject's value for CLASS2 should be JUN (for "junior").
  - If a given subject's value for CLASS is 4, then that subject's value for CLASS2 should be SEN (for "senior").

4. Following this second DATA step, use PROC PRINT again to print out all of the raw data included in the new data set, D2 (all variables).
5. Submit the program for analysis, and, if necessary, correct it so that it runs without errors.
6. Review your log window and output files for possible errors.

What You Will Hand In:

Hand in the following materials stapled together in the following order:

1. A printout of your SAS program.
2. A printout of your SAS log.
3. A printout of your SAS output.

Hint Regarding the SAS Log Window

The following is a short excerpt from the SAS log window (the actual log is much longer). This note appears in the SAS log after the data manipulation statements, but before the final PROC PRINT statements. This note provides information about the generation of missing values. It is included here to remind you that a note concerning the generation of missing values is not necessarily a cause for alarm.

If your SAS program ran correctly, you should receive a note about missing values similar to the following note. However, the note you receive will not necessarily be identical to the following note with respect to the line numbers that are listed.

```
NOTE: Missing values were generated as a result of performing an operation on
      missing values.
      Each place is given by: (Number of times) at (Line):(Column).
      1 at 32:11   1 at 34:16   2 at 34:21   2 at 34:26   2 at 34:32
```

Hint Regarding the SAS Output

If your program ran correctly, the results from the first PROC PRINT in your initial data set (D1) should resemble the following:

```
                              JOHN DOE                                   1
  Obs     SUB_NUM     Q1     Q2     Q3     Q4     SEX     AGE     CLASS

    1        1         6      .      2      7      F       20       1
    2        2         3      2      7      2      M       26       1
    3        3         7      7      1      7      M       19       1
    4        4         5      6      .      5      F       23       2
    5        5         6      7      1      6      M       21       2
    6        6         3      2      6      3      F       25       2
    7        7         5      6      2      5      F       25       3
    8        8         5      6      1      5      F       23       3
    9        9         7      7      1      6      M       31       3
   10       10         5      4      1      4      M       25       4
   11       11         4      5      3      5      F       42       4
   12       12         7      6      1      6      F       38       4
```

If your program ran correctly, the results from the second PROC PRINT in your second data set (D2) should resemble the following:

```
                              JOHN DOE                                    2

Obs   SUB_NUM  Q1   Q2   Q3   Q4   SEX   AGE   CLASS   DEVEL   AGE2   CLASS2

  1       1     6    .    6    7    F     20     1        .       0     FRE
  2       2     3    2    1    2    M     26     1      2.00      0     FRE
  3       3     7    7    7    7    M     19     1      7.00      0     FRE
  4       4     5    6    .    5    F     23     2        .       0     SOP
  5       5     6    7    7    6    M     21     2      6.50      0     SOP
  6       6     3    2    2    3    F     25     2      2.50      0     SOP
  7       7     5    6    6    5    F     25     3      5.50      0     JUN
  8       8     5    6    7    5    F     23     3      5.75      0     JUN
  9       9     7    7    7    6    M     31     3      6.75      1     JUN
 10      10     5    4    7    4    M     25     4      5.00      0     SEN
 11      11     4    5    5    5    F     42     4      4.75      1     SEN
 12      12     7    6    7    6    F     38     4      6.50      1     SEN
```

Exercise 8.2: Using Subsetting IF Statements with the Academic Development Data

Overview

In this exercise you will analyze data from the fictitious academic development study first presented in Exercise 8.1. The SAS program that you write will create a new data set that contains data from the freshmen in your sample study, and it will then perform PROC MEANS on that new data set. The program will then go on to repeat this process for each group of sophomores, juniors, and seniors. Finally, your program will create a data set that contains data from just those subjects who had no missing data for variables Q1, Q2, Q3, and Q4, and it will then perform PROC PRINT on the new data set. All of this will be done within one SAS program.

The Study and the Data Set to be Analyzed

The initial DATA step for this exercise should be identical to the initial DATA step used with Exercise 8.1, in which you analyzed the data set that appeared in Table 8.E1.1. If you completed that exercise and saved your SAS program as a computer file, you should open that file and delete everything except for the initial DATA step. This means that you should keep only the initial OPTIONS statement, DATA statement, INPUT statement, DATALINES statement, and data lines. You should delete the data manipulation statements that were unique to Exercise 8.1. After you have deleted the old data manipulation statements, you can then append the new data manipulation statements that are required for the current exercise.

If for any reason you do not have the DATA step from Exercise 8.1 available, you will have to prepare it at this time. To do this, follow the instructions provided in Exercise 8.1 in the sections titled "The Study," "Data Set to be Analyzed," and step 1 of "Your Assignment."

Your Assignment

In this exercise, you will write a single SAS program that performs a number of data subsetting tasks. This program will also include PROC MEANS statements that will compute descriptive statistics and a PROC PRINT statement that will print the raw data from the data sets that you create. The most important thing to remember is that all of the following tasks must be performed within a single SAS program.

Use the SAS programs presented in Chapter 8 of the *Student Guide* as models.

In the SAS program that you write, do the following:

1. Create a SAS program that will input the data set (D1) that is presented in Table 8.E1.1 in Exercise 8.1, as previously described.
2. Following the initial DATA step, create a new data set titled "D2," and make it a duplicate of D1. Subjects should be retained in the new data set only if their value for the variable CLASS is equal to 1, which means that only freshmen will be retained in this new data set.
    - Perform PROC MEANS on the new data set that you have just created.
    - Have the TITLE1 statement following this PROC MEANS read:

      ```
      TITLE1  'your name --FRESHMEN';
      ```

      Use your actual name in the place of "your name" in the preceding TITLE1 statement.
3. Create a new data set titled "D3," and make it a duplicate of D1. Subjects should be retained in the new data set only if their value for the variable CLASS is equal to 2, which means that only sophomores will be retained in this new data set.
    - Perform PROC MEANS on the new data set that you have just created.
    - Have the TITLE1 statement following this PROC MEANS read:

      ```
      TITLE1  'your name --SOPHOMORES';
      ```
4. Create a new data set titled "D4," and make it a duplicate of D1. Subjects should be retained in the new data set only if their value for the variable

CLASS is equal to 3, which means that only juniors will be retained in this new data set.

- Perform PROC MEANS on the new data set that you have just created.
- Have the TITLE1 statement following this PROC MEANS read:

```
TITLE1  'your name --JUNIORS';
```

5. Create a new data set titled "D5," and make it a duplicate of D1. Subjects should be retained in the new data set only if their value for the variable CLASS is equal to 4, which means that only seniors will be retained in this new data set.
   - Perform PROC MEANS on the new data set that you have just created.
   - Have the TITLE1 statement following this PROC MEANS read:

   ```
   TITLE1  'your name --SENIORS';
   ```

6. Create a new data set titled "D6," and make it a duplicate of D1. Subjects should be retained in the new data set only if they have no missing data for Q1, Q2, Q3, and Q4.
   - Perform PROC PRINT on the new data set that you have just created.
   - Use the VAR statement with this PROC PRINT so that raw scores are printed only for the variables Q1, Q2, Q3, and Q4.
   - Include your name in the TITLE1 statement only.
7. Submit your SAS program for analysis, and, if necessary, correct it so that it runs without errors.

What You Will Hand In

Hand in the following materials stapled together in the following order:

1. A printout of your SAS program.
2. A printout of your SAS log.
3. A printout of your SAS output.

Hint

If your program ran correctly, your output should resemble the following:

```
                           JOHN DOE--FRESHMEN                              1

                          The MEANS Procedure

Variable     N          Mean       Std Dev       Minimum       Maximum
-----------------------------------------------------------------------
SUB_NUM      3     2.0000000     1.0000000     1.0000000     3.0000000
Q1           3     5.3333333     2.0816660     3.0000000     7.0000000
Q2           2     4.5000000     3.5355339     2.0000000     7.0000000
Q3           3     3.3333333     3.2145503     1.0000000     7.0000000
Q4           3     5.3333333     2.8867513     2.0000000     7.0000000
AGE          3    21.6666667     3.7859389    19.0000000    26.0000000
CLASS        3     1.0000000             0     1.0000000     1.0000000
-----------------------------------------------------------------------
```

```
                           JOHN DOE--SOPHOMORES                            2

                           The MEANS Procedure

Variable     N          Mean       Std Dev       Minimum       Maximum
-----------------------------------------------------------------------
SUB_NUM      3     5.0000000     1.0000000     4.0000000     6.0000000
Q1           3     4.6666667     1.5275252     3.0000000     6.0000000
Q2           3     5.0000000     2.6457513     2.0000000     7.0000000
Q3           2     3.5000000     3.5355339     1.0000000     6.0000000
Q4           3     4.6666667     1.5275252     3.0000000     6.0000000
AGE          3    23.0000000     2.0000000    21.0000000    25.0000000
CLASS        3     2.0000000             0     2.0000000     2.0000000
-----------------------------------------------------------------------
```

```
                              JOHN DOE--JUNIORS                              3

                             The MEANS Procedure

Variable      N           Mean        Std Dev        Minimum        Maximum
-----------------------------------------------------------------------------
SUB_NUM       3      8.0000000      1.0000000      7.0000000      9.0000000
Q1            3      5.6666667      1.1547005      5.0000000      7.0000000
Q2            3      6.3333333      0.5773503      6.0000000      7.0000000
Q3            3      1.3333333      0.5773503      1.0000000      2.0000000
Q4            3      5.3333333      0.5773503      5.0000000      6.0000000
AGE           3     26.3333333      4.1633320     23.0000000     31.0000000
CLASS         3      3.0000000              0      3.0000000      3.0000000
-----------------------------------------------------------------------------
```

```
                             JOHN DOE--SENIORS                               4

                            The MEANS Procedure

Variable      N           Mean        Std Dev        Minimum        Maximum
-----------------------------------------------------------------------------
SUB_NUM       3     11.0000000      1.0000000     10.0000000     12.0000000
Q1            3      5.3333333      1.5275252      4.0000000      7.0000000
Q2            3      5.0000000      1.0000000      4.0000000      6.0000000
Q3            3      1.6666667      1.1547005      1.0000000      3.0000000
Q4            3      5.0000000      1.0000000      4.0000000      6.0000000
AGE           3     35.0000000      8.8881944     25.0000000     42.0000000
CLASS         3      4.0000000              0      4.0000000      4.0000000
-----------------------------------------------------------------------------
```

JOHN DOE 5

Obs	Q1	Q2	Q3	Q4
1	3	2	7	2
2	7	7	1	7
3	6	7	1	6
4	3	2	6	3
5	5	6	2	5
6	5	6	1	5
7	7	7	1	6
8	5	4	1	4
9	4	5	3	5
10	7	6	1	6

Exercises for Chapter 9: *z* Scores

Exercise 9.1: Satisfaction with Academic Development and the Social Environment Among College Students

Overview

In this exercise, you will create a SAS data set that contains raw scores on two variables for 12 students. You will use PROC MEANS to compute the means and sample standard deviations for these variables. You will then use data manipulation statements to convert the raw-score variables into *z* score variables. You will perform PROC PRINT and PROC MEANS on the new data set that contains the *z* score variables. You will do all of the preceding steps within a single SAS program. Finally, you will answer a number of questions about the resulting SAS output to demonstrate that you are able to correctly interpret *z* scores.

The Study

Suppose that you are conducting research in which you assess the satisfaction of 12 students (subjects) with various aspects of their college experience. You are interested in assessing their satisfaction in the following two areas:

- Satisfaction with academic development (the extent to which they are satisfied with their intellectual growth and scholarly performance).

- Satisfaction with the social environment (the extent to which they are satisfied with their opportunities for making friends and interacting with peers).

You develop a questionnaire to assess these two variables as follows:

The academic development scale assesses satisfaction with academic development. This scale consists of four statements such as "I am satisfied with the extent of my academic development since enrolling here." Subjects respond using a 7-point response format in which "1" = "Disagree Very Strongly" and "7" = "Agree Very Strongly." To create a single academic development score for a given subject, you add together that subject's responses to the four items constituting the academic development scale. Scores on this academic development scale can therefore range from a low of 4 (if the subject circles "1" for "Disagree Very Strongly" for each item) to a high of 28 (if the subject circles "7" for "Agree Very Strongly" for each item). For this variable, higher scores indicate higher levels of satisfaction with academic development.

The social environment scale assesses satisfaction with the social environment. This scale consists of six statements such as "I am satisfied with the social environment at this college." Students respond using a 7-point response format in which "1" = "Disagree Very Strongly" and "7" = "Agree Very Strongly." To create a single social environment score for a given subject, you add together that subject's responses to the six items constituting the social environment scale. Scores on this social environment scale can therefore range from a low of 6 (if the subject circles "1" for "Disagree Very Strongly" for each item) to a high of 42 (if the subject circles "7" for "Agree Very Strongly" for each item). For this variable, higher scores indicate higher levels of satisfaction with the social environment.

Data Set to be Analyzed

Suppose that you administer the questionnaire to 12 college students (subjects), and compute each subject's score on both scales. Table 9.E1.1 presents these scores.

Table 9.E1.1

Subjects' Scores on the Academic Development Scale and the Social Environment Scale

Subject	Academic development	Social environment
01. Fred	25	34
02. Susan	16	30
03 Marsha	4	42
04. Charles	12	24
05. Paul	28	39
06. Cindy	15	29
07. Jack	21	27
08. Cathy	8	36
09. George	23	21
10. John	6	32
11. Marie	10	19
12. Emmett	19	33

Overview of the Analysis

You will write a single SAS program that will perform all of the tasks described in this exercise, which is divided into Step 1 and Step 2 (as was done in the corresponding chapter in the *Student Guide*). You will first write a SAS program to complete the tasks assigned in Step 1. After you have reviewed the output produced in Step 1, you will then add new SAS statements to the same program to complete the tasks associated with Step 2. After you have produced output from the complete SAS program, you will answer a number of questions pertaining to that output.

Your Assignment Regarding Step 1 of the Analysis

1. Write a SAS program that will input and analyze the data presented in Table 9.E1.1. When you write this program, do the following:
   - Use the SAS program presented in Chapter 9 of the *Student Guide* as a model to create a data set named "D1."
   - Include subject numbers as a variable in your data set (as shown in the program presented in the *Student Guide*). Use the SAS variable name SUB_NUM to represent this variable.
   - Include subject names as a variable in your data set. Use the SAS variable name NAME to represent this variable.
   - Use the SAS variable name ACADEM to represent subject scores on the academic development scale.
   - Use the SAS variable name SOCIAL to represent subject scores on the social environment scale.
   - Perform PROC MEANS on ACADEM and SOCIAL so that you will be able to review the means and sample standard deviations for these variables (along with the other appropriate descriptive statistics).
   - Type your full name in the TITLE1 statement, so that it will appear in the output.
2. Submit the program for analysis, and, if necessary, correct it so that it runs without errors.

Your Assignment Regarding Step 2 of the Analysis

Remember that the SAS statements that you write in Step 2 will be appended to the SAS statements that you wrote for Step 1. All of the following tasks should be included within a single SAS program.

1. Begin a new DATA step in your SAS program by creating a new data set. Name the new data set "D2," and initially create it as a duplicate of existing data set D1.
2. Within this new DATA step, create a new variable named ACADEM_Z. This variable should be the *z*-score variable that corresponds to the raw-

score variable ACADEM (from Step 1). A given subject's score on ACADEM_Z should be the *z* score that corresponds to his or her raw score for ACADEM. You will write a data manipulation statement to create this *z*-score variable. You will have to review the output of the PROC MEANS from Step 1 to find the mean and standard deviation to include in this data manipulation statement.

3. Within the same DATA step, create a second new variable named SOCIAL_Z. This variable should be the *z*-score variable that corresponds to the raw-score variable SOCIAL (from Step 1). A given subject's score for SOCIAL_Z should be the *z* score that corresponds to his or her raw score for SOCIAL. You will write a data manipulation statement to create this *z*-score variable. You will have to review the output of the PROC MEANS from Step 1 to find the mean and standard deviation that must be included in this data manipulation statement.
4. Use PROC PRINT to create output that lists each subject's score for the following variables in this order: NAME, ACADEM, SOCIAL, ACADEM_Z, and SOCIAL_Z.
5. Use PROC MEANS to compute the mean, sample standard deviation, and other descriptive statistics for the variables ACADEM_Z and SOCIAL_Z.
6. Submit the program for analysis, and, if necessary, correct it so that it runs without errors.

Questions Regarding the SAS Output

On a new sheet of paper, write your full name and the title, "Exercise 9.1: Answers to Questions." On this sheet, answer the following questions. You do not have to rewrite each of these questions on your sheet, but make sure that you number your answers. Please type or print your answers (do not use cursive).

1. **Question**: The third SAS procedure in your program should have performed PROC MEANS on ACADEM_Z and SOCIAL_Z. Based on the means and standard deviations for these *z*-score variables, is there reason to believe that the *z*-score variables were created correctly? Explain your answer.

Hint: For help with answering this question, see the section titled "Reviewing the mean and standard deviation for the new *z*-score variable," which appears near Output 9.3 in the *Student Guide*.

The remaining questions are based on the results of PROC PRINT, as they appear in your SAS output.

2. **Question**: Marsha's raw score on ACADEM was 4 (Marsha was observation 3). What was the relative standing of this score within the sample? Explain your answer. (Hint: Your answer should refer to the *z* score that corresponds to this raw score as stated in Chapter 9 of the *Student Guide*).
3. **Question**: Jack's raw score on ACADEM was 21 (Jack was observation 7). What was the relative standing of this score within the sample? Explain your answer.
4. **Question**: Compared to the other subjects, did Fred (observation 1) score higher on the academic development scale or on the social environment scale? Explain your answer. (Hint: Your answer should refer to the *z* scores for these variables as stated in Chapter 9 of the *Student Guide*).
5. **Question**: Compared to the other subjects, did Marie (observation 11) score higher on the academic development scale or on the social environment scale? Explain your answer.
6. **Question**: Compared to the other subjects, did Susan (observation 2) score higher on the academic development scale or on the social environment scale? Explain your answer.

What You Will Hand In

Hand in the following materials stapled together in this order.

1. A printout of your SAS program (including data), your SAS log, and SAS output files.
2. Your sheet of paper titled "Exercise 9.1: Answers to Questions" on which you wrote answers to the preceding questions.

Hint

If your SAS program ran correctly, the results of your PROC PRINT should resemble the following output. Remember that your program will produce more output, in addition to the output reproduced below.

```
                           JOHN DOE

 Obs    NAME       ACADEM    SOCIAL    ACADEM_Z    SOCIAL_Z

   1    Fred         25        34       1.26443     0.52395
   2    Susan        16        30       0.05638    -0.07485
   3    Marsha        4        42      -1.55436     1.72156
   4    Charles      12        24      -0.48054    -0.97305
   5    Paul         28        39       1.66711     1.27246
   6    Cindy        15        29      -0.07785    -0.22455
   7    Jack         21        27       0.72752    -0.52395
   8    Cathy         8        36      -1.01745     0.82335
   9    George       23        21       0.99597    -1.42216
  10    John          6        32      -1.28591     0.22455
  11    Marie        10        19      -0.74899    -1.72156
  12    Emmett       19        33       0.45906     0.37425
```

Exercise 9.2: Physical Fitness and Verbal Ability

Overview

In this exercise, you will create a SAS data set that contains raw scores on two variables for 12 students. You will use PROC MEANS to compute the means and sample standard deviations for these variables. You will then use data manipulation statements to convert the raw-score variables into *z* score variables. You will perform PROC PRINT and PROC MEANS on the new data set that contains the *z* score variables. Finally, you will answer a number of questions about the resulting SAS output to demonstrate that you are able to correctly interpret *z* scores.

The Study

Suppose that you are conducting research in which you obtain measures on two very different variables. From a sample study, you obtain scores for 12 students (subjects) based on the following two variables:

Physical fitness. To assess physical fitness, you ask the subjects to perform a variety of physical tasks such as doing push-ups, doing sit-ups, running 1 kilometer, and so forth. They earn points according to (a) the number of repetitions they can complete (for example, the number of push-ups), and (b) the speed with which they can complete other tasks (for example, their elapsed time in running 1 kilometer). For each subject, you add together the points earned on each task to create a single score that represents their level of physical fitness. Scores can range from 0 to 500, with higher scores representing higher levels of fitness.

Verbal ability. To assess verbal ability, you administer an 80-item test to the subjects. The test consists of multiple-choice items that assess vocabulary, reading comprehension, and other verbal skills. Scores on the test can range from 0 to 80, with higher scores representing higher levels of verbal ability.

Data Set to be Analyzed

Suppose that you administer the physical fitness test and verbal ability test to 12 subjects. Table 9.E2.1 presents their scores.

Table 9.E2.1

Scores on the Physical Fitness and Verbal Ability Variables

Subject	Physical fitness	Verbal ability
01. Fred	300	40
02. Susan	240	50
03. Marsha	350	30
04. Charles	250	55
05. Paul	260	25
06. Cindy	150	45
07. Jack	270	70
08. Cathy	200	65
09. George	230	20
10. John	190	35
11. Marie	380	50
12. Emmett	240	35

Overview of the Analysis

You will write a single SAS program that will perform all of the tasks described in this exercise, which is divided into Step 1 and Step 2 (as was done in the corresponding chapter in the *Student Guide*). You will first write a SAS program to complete the tasks assigned in Step 1. After you have reviewed the output produced in Step 1, you will then add new SAS statements to the same program to complete the tasks associated with Step 2. After you have produced output from the complete SAS program, you will answer a number of questions pertaining to that output.

Your Assignment Regarding Step 1 of the Analysis

1. Write a SAS program that will input and analyze the data presented in Table 9.E2.1. When you write this program, do the following:
   - Use the SAS program presented in Chapter 9 of the *Student Guide* as a model to create a data set named "D1."
   - Include subject numbers as a variable in your data set (as shown in the program presented in the *Student Guide*). Use the SAS variable name SUB_NUM to represent this variable.
   - Include subject names as a variable in your data set. Use the SAS variable name NAME to represent this variable.
   - Use the SAS variable name FIT to represent subject scores on the physical fitness variable.
   - Use the SAS variable name VERBAL to represent subject scores on the verbal ability test.
   - Perform PROC MEANS on FIT and VERBAL so that you will be able to review the means and sample standard deviations for these variables (along with the other appropriate descriptive statistics).
   - Type your full name in the TITLE1 statement, so that it will appear in the output.
2. Submit the program for analysis, and, if necessary, correct it so that it runs without errors.

Your Assignment Regarding Step 2 of the Analysis

Remember that the SAS statements that you write in Step 2 will be appended to the SAS statements that you wrote for Step 1. All of the following tasks should be included within a single SAS program.

1. Begin a new DATA step in your SAS program by creating a new data set. Name the new data set "D2," and initially create it as a duplicate of existing data set D1.

2. Within this new DATA step, create a new variable named FIT_Z. This variable should be the *z*-score variable that corresponds to the raw-score variable FIT (from Step 1). A given subject's score on FIT_Z should be the *z* score that corresponds to his or her raw score for FIT. You will write a data manipulation statement to create this *z*-score variable. You will have to review the output of the PROC MEANS from Step 1 to find the mean and standard deviation to include in this data manipulation statement.
3. Within the same DATA step, create a second new variable named VERBAL_Z. This variable should be the *z*-score variable that corresponds to the raw-score variable VERBAL (from Step 1). A given subject's score for VERBAL_Z should be the *z* score that corresponds to his or her raw score for VERBAL. You will write a data manipulation statement to create this *z*-score variable. You will have to review the output of the PROC MEANS from Step 1 to find the mean and standard deviation to include in this data manipulation statement.
4. Use PROC PRINT to create output that lists each subject's score for the following variables in this order: NAME, FIT, VERBAL, FIT_Z, and VERBAL_Z.
5. Use PROC MEANS to compute the mean, sample standard deviation, and other descriptive statistics for the variables FIT_Z and VERBAL_Z.
6. Submit the program for analysis, and, if necessary, correct it so that it runs without errors.

Questions Regarding the SAS Output

On a new sheet of paper, write your full name, and the title, "Exercise 9.2: Answers to Questions." On this sheet, answer the following questions. You do not have to rewrite each of these questions on your sheet, but make sure that you number your answers. Please type or print your answers (do not use cursive).

1. **Question**: The third SAS procedure in your program should have performed PROC MEANS on FIT_Z and VERBAL_Z. Based on the means and standard deviations for these *z*-score variables, is there reason

to believe that the *z*-score variables were created correctly? Explain your answer.

Hint: For help with answering this question, see the section titled "Reviewing the mean and standard deviation for the new *z*-score variable," which appears near Output 9.3 in the *Student Guide*.

The remaining questions are based on the results of PROC PRINT, as they appear in your SAS output.

2. **Question**: Marie's raw score on FIT was 380 (Marie was observation 11). What was the relative standing of this score within the sample? Explain your answer. (Hint: Your answer should refer to the *z* score that corresponds to this raw score as stated in Chapter 9 of the *Student Guide*).

3. **Question**: Susan's raw score on FIT was 240 (Susan was observation 2). What was the relative standing of this score within the sample? Explain your answer.

4. **Question**: Compared to the other subjects, did Paul (observation 5) score higher on the physical fitness test or on the verbal ability test? Explain your answer. (Hint: Your answer should refer to the *z* scores for these variables as stated in Chapter 9 of the *Student Guide*).

5. **Question**: Compared to the other subjects, did George (observation 9) score higher on the physical fitness test or on the verbal ability test? Explain your answer.

6. **Question**: Compared to the other subjects, did Jack (observation 7) score higher on the physical fitness test or on the verbal ability test? Explain your answer.

What You Will Hand In

Hand in the following materials stapled together in this order.

1. A printout of your SAS program (including data), your SAS log, and SAS output files.
2. Your sheet of paper titled "Exercise 9.2: Answers to Questions" on which you wrote answers to the preceding questions.

Hint

If your SAS program ran correctly, the results of your PROC PRINT should resemble the following output. Remember that your program will produce more output than is shown here.

```
                    JANE DOE                                2

Obs      NAME        FIT    VERBAL       FIT_Z      VERBAL_Z

  1      Fred        300      40        0.72440     -0.22546
  2      Susan       240      50       -0.24147      0.45159
  3      Marsha      350      30        1.52930     -0.90251
  4      Charles     250      55       -0.08049      0.79012
  5      Paul        260      25        0.08049     -1.24103
  6      Cindy       150      45       -1.69028      0.11307
  7      Jack        270      70        0.24147      1.80569
  8      Cathy       200      65       -0.88538      1.46716
  9      George      230      20       -0.40245     -1.57955
 10      John        190      35       -1.04636     -0.56398
 11      Marie       380      50        2.01223      0.45159
 12      Emmett      240      35       -0.24147     -0.56398
```

EXERCISES 10

Exercises for Chapter 10: Bivariate Correlation

Exercise 10.1: Correlational Study of Drinking and Driving Behavior

Overview

What attitudinal variables are correlated with the behavior of drinking and driving? Are people less likely to drink and drive if they are morally committed to the legal norm? Are they less likely to drink and drive if they believe that they are likely to be arrested? The current study was designed to answer these questions.

In this exercise, you will analyze data from two time periods and determine whether moral commitment to the legal norm is correlated with the drinking and driving behavior of subjects at Time 2. You will use PROC PLOT to create a scattergram for your study variables, and you will use PROC CORR to compute the Pearson correlation between them. In addition, you will prepare a report that summarizes the results of your analysis.

Note: Although the study reported here is fictitious, it was inspired by the actual study reported by Green (1991).

The Study

Suppose that you conduct a correlational study on a group of 21 adults (subjects) who sometimes drink alcoholic beverages. Your study takes place during the months of June and July one summer, and you obtain data at two points in time.

Your study includes three predictor variables: moral commitment to the legal norm, perceived certainty of arrest, and drinking and driving behavior at the end of June (Time 1). The single criterion variable in your study is drinking and driving behavior at the end of July (Time 2). From your sample study, you obtain scores for 21 subjects based on these variables as follows:

Moral commitment to the legal norm. The legal norm holds that driving a car while drinking alcohol is wrong. Therefore, people who score high on moral commitment to the legal norm will tend to agree that drinking and driving is wrong.

Suppose that you measure "moral commitment" using a 4-item scale. The scale consists of statements such as "It is wrong to drive a car while under the influence of alcohol." Subjects respond to these statements using a 7-point scale in which "1" represents "Disagree Very Strongly" and "7" represents "Agree Very Strongly." For each subject, you sum responses to the four items to construct a single measure of moral commitment that can range from a low score of 4 to a high score of 28. With this measure, higher scores represent a higher level of commitment to the legal norm.

When analyzing your data, use the SAS variable name "MC_T1" to represent this variable. Here, MC_T1 stands for "moral commitment at Time 1."

Perceived certainty of arrest. Assume that you also use a 4-item scale to measure the subjects' belief that they will be arrested if they drink and drive. The scale consists of statements such as "There is a good chance I will be arrested if I drive a car under the influence of alcohol." Again, they respond to these statements using a 7-point scale in which "1" represents "Disagree Very Strongly" and "7" represents "Agree Very Strongly." You sum responses to the four items to construct a single measure of perceived certainty. This variable can range from a low score of 4 to a high score of 28. With this measure, higher scores represent greater certainty that they will be arrested if they drink and drive.

When analyzing your data, use the SAS variable name "PC_T1" to represent this variable. Here, PC_T1 stands for "perceived certainty at Time 1."

Drinking and driving at Time 1. At the end of the month of June, you administer the previously mentioned scales that assess moral commitment to the legal norm and perceived certainty of arrest. You also ask subjects to indicate the number of times that they drank and drove during the month of June. When you analyze your data, you will determine whether this variable is correlated with the number of times that they drank and drove during the following month (July).

When analyzing your data, use the SAS variable name "DD_T1" to represent this variable. Here, DD_T1 stands for "drinking and driving at Time 1."

Drinking and driving at Time 2. At the end of the following month (July), you ask subjects to again indicate the number of times that they drank and drove during that month. This measure will serve as the primary criterion variable that you are trying to predict.

When analyzing your data, use the SAS variable name "DD_T2" to represent this variable. Here, DD_T2 stands for "drinking and driving at Time 2."

Data Set to be Analyzed

Table 10.E1.1 presents fictitious scores for each of the 21 subjects on each of the four variables in this study.

Table 10.E1.1

Variables Analyzed in the Drinking and Driving Study

Subject	Moral commitment to the legal norm	Perceived certainty of arrest	Drinking and driving at Time 1	Drinking and driving at Time 2
01	24	8	1	3
02	12	12	3	4
03	4	4	4	5
04	16	12	2	2
05	28	.	1	1
06	8	16	3	3
07	12	8	3	5
08	8	8	5	6
09	16	16	4	4
10	16	16	4	5
11	12	12	4	6
12	20	16	5	2
13	28	24	5	3
14	20	20	5	5
15	20	24	6	4
16	16	20	6	6
17	16	16	6	7
18	4	12	6	8
19	24	28	7	5
20	20	24	7	6
21	4	20	8	8

Table 10.E1.1 uses the standard format in which the rows represent the subjects, and the columns represent the variables. You can see that the first subject displays

- a Subject number of 01
- a score of 24 on Moral commitment to the legal norm
- a score of 8 on Perceived certainty of arrest
- a score of 1 on Drinking and driving at Time 1
- a score of 3 on Drinking and driving at Time 2.

Data for the remaining subjects can be interpreted in the same way. Note that Subject 05 has missing data on perceived certainty. Be sure that you actually type the period that appears in this location.

Your Assignment Regarding the SAS Program

1. Write a SAS program that will input and analyze the data set presented in Table 10.E1.1. When you write the DATA step of this program, do the following:
   - Use the SAS programs presented in Chapter 10 of the *Student Guide* as models.
   - Type all of the information from Table 10.E1.1. You should type the Subject number variable as the first column as it appears in the table. Name this variable SUB_NUM. When you type this SAS variable name, make sure you use an underscore (_) and not a hyphen (-). When you type subject numbers such as "01," make sure you type a zero (0) and not the letter O.
   - In the DATA step, remember to use the SAS variable names MC_T1, PC_T1, DD_T1, and DD_T2 to represent the four data variables.
2. When you write the PROC step of this program, do the following:
   - Use PROC PLOT to create a scattergram in which the criterion variable is drinking and driving behavior at Time 2, and the predictor variable is moral commitment to the legal norm.
   - Type your full name in the TITLE1 statement following this PROC step, so that it will appear in the output.
   - Use PROC CORR to compute all possible correlations between all four data variables: MC_T1, PC_T1, DD_T1, and DD_T2. Do not include the variable SUB_NUM in this correlation matrix. This should be done using a single PROC CORR statement. For more information, see the section in Chapter 10 titled "Using PROC CORR to Compute All Possible Correlations for a Group of Variables."
3. Submit the program for analysis, and, if necessary, correct it so that it runs without errors.

4. Assume that you began with the following research question: "The purpose of your study was to determine whether moral commitment to the legal norm is correlated with the drinking and driving behavior of subjects at Time 2." Your research hypothesis is that there will be a negative relationship between moral commitment to the legal norm and drinking and driving behavior. One of the correlations produced by the PROC CORR statement in your program is relevant to answering this question.

   Prepare a report summarizing the results of your analysis using the format shown in the section titled "Summarizing the Results of the Analysis" from Chapter 10 of the *Student Guide*. You will need to modify the report so that it is relevant to the analysis that you are conducting in this exercise. Underline each heading in this report, and do not forget to include information for sections A-L.

   Although your PROC CORR statement produced a number of correlations, you are to focus on only one for this exercise—the correlation between the following:

   (a) moral commitment to the legal norm

   (b) drinking and driving behavior at Time 2.

What You Will Hand In

Hand in the following materials stapled together in this order:

1. A printout of your SAS program, your SAS log, and your SAS output files.

   Your SAS output should include both

   (a) the scattergram in which the criterion variable was drinking and driving behavior at Time 2 and the predictor variable was moral commitment to the legal norm (you do not have to draw a "best fitting line" on this scattergram)

   (b) the results of the PROC CORR statement that computed all possible correlations between the four data variables.

2. Your report summarizing the results of the analysis, including information for sections A-L from Chapter 10 of the *Student Guide.*

Hint

If your SAS program ran correctly, your output should resemble the following output:

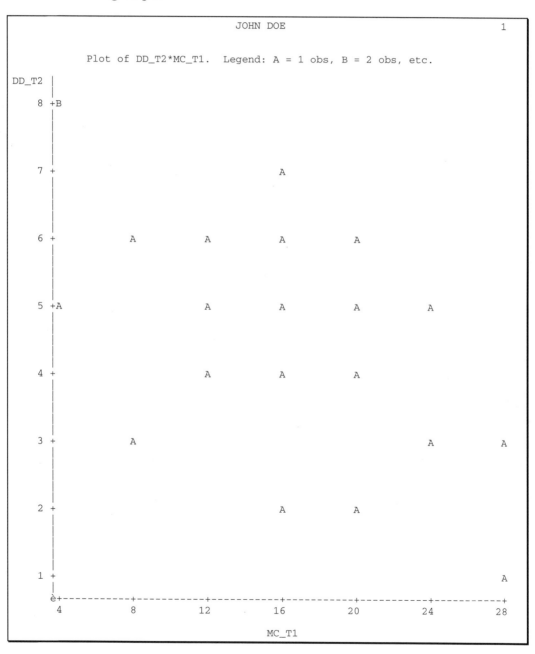

```
                                JOHN DOE                                 1

            Plot of DD_T2*MC_T1.  Legend: A = 1 obs, B = 2 obs, etc.

DD_T2 |
      |
    8 +B
      |
      |
      |
      |
    7 +                                    A
      |
      |
      |
      |
    6 +            A           A           A           A
      |
      |
      |
      |
    5 +A                       A           A           A           A
      |
      |
      |
      |
    4 +                        A           A           A
      |
      |
      |
      |
    3 +            A                                               A           A
      |
      |
      |
      |
    2 +                                    A           A
      |
      |
      |
      |
    1 +                                                                        A
      |
      è+-----------+-----------+-----------+-----------+-----------+-----------+
       4           8          12          16          20          24          28

                                         MC_T1
```

JOHN DOE 2

The CORR Procedure

4 Variables: MC_T1 PC_T1 DD_T1 DD_T2

Simple Statistics

Variable	N	Mean	Std Dev	Sum	Minimum	Maximum
MC_T1	21	15.61905	7.36530	328.00000	4.00000	28.00000
PC_T1	20	15.80000	6.42036	316.00000	4.00000	28.00000
DD_T1	21	4.52381	1.91361	95.00000	1.00000	8.00000
DD_T2	21	4.66667	1.90613	98.00000	1.00000	8.00000

Pearson Correlation Coefficients
Prob > |r| under H0: Rho=0
Number of Observations

	MC_T1	PC_T1	DD_T1	DD_T2
MC_T1	1.00000	0.55954	-0.19799	-0.59358
		0.0103	0.3896	0.0046
	21	20	21	21
PC_T1	0.55954	1.00000	0.65765	0.03456
	0.0103		0.0016	0.8850
	20	20	20	20
DD_T1	-0.19799	0.65765	1.00000	0.70823
	0.3896	0.0016		0.0003
	21	20	21	21
DD_T2	-0.59358	0.03456	0.70823	1.00000
	0.0046	0.8850	0.0003	
	21	20	21	21

Reference

Green, D. E. (1991). Inhibition, motivation, and self-reported involvement in drinking and driving behavior. *Criminal Justice Review, 16,* 1-16.

Exercise 10.2: Correlational Study of Nurses' Intent to Remain

Overview

People who conduct research in organizations often measure employees' intent to remain with the organization. Employers generally try to hire and retain employees who intend to remain with them for a long time, because it is often expensive to replace employees.

The present study investigates variables that may be related to the intent to remain in a job using a sample of nurses. In this exercise, you will focus on the relationship between the nurses' intent to remain and the nurses' assertiveness. You will use PROC PLOT to create a scattergram for these variables and will use PROC CORR to compute the Pearson correlation between them. Finally, you will prepare a report that summarizes the results of your analysis.

Note: Although the study reported here is fictitious, it was inspired by the actual study reported by Ellis and Miller (1993).

The Study

Suppose that you conduct a study to investigate variables that may be correlated with nurses' intent to remain with their current employers. You develop a questionnaire that assesses their intent to remain, as well as two variables that you believe may affect their intent to remain: (a) nurse assertiveness and (b) emotional exhaustion.

You administer the questionnaire to 22 nurses (subjects) who are currently employed in hospitals and similar settings.

Your study includes two predictor variables: nurse assertiveness and emotional exhaustion. The single criterion variable is the nurses' intent to remain with their current employer. From your sample study, you obtain scores for 22 subjects based on these variables as follows:

Nurse assertiveness. Nurse assertiveness refers to the extent to which nurses speak up, voice their opinions, express their sense of self-worth, and validate their own professional status (particularly in interactions with physicians). Suppose that you assess nurse assertiveness using a 5-item scale. The scale consists of statements such as "I tell physicians when, in my opinion, the approach they are using to treat a patient is not likely to be effective." Subjects respond to these statements using a 7-point scale in which "1" represents "Disagree Very Strongly" and "7" represents "Agree Very Strongly." For each subject, you sum responses to the five items to construct a single measure of nurse assertiveness that can range from a low score of 5 to a high score of 35. With this measure, higher scores represent a higher level of nurse assertiveness. When analyzing your data, use the SAS variable name "N_ASSERT" to represent this variable.

Emotional exhaustion. Emotional exhaustion refers to the extent to which nurses are psychologically fatigued from the stress and other problems experienced in their jobs. Suppose that you assess emotional exhaustion using a 5-item scale. The scale consists of statements such as "I feel tired when I think about having to face another day at work." Subjects respond to these statements using the same 7-point "Agree-Disagree" format previously described. For each subject, you sum responses to the five items to construct a single measure of emotional exhaustion that can range from a low score of 5 to a high score of 35. With this measure, higher scores represent a higher level of emotional exhaustion. When analyzing your data, use the SAS variable name "EXHAUST" to represent this variable.

Intent to remain. Intent to remain refers to the extent to which nurses plan to remain employed with their current organizations. Suppose that you assess intent to remain using a 5-item scale. The scale consists of statements such as "I plan to remain employed at this hospital for a long time." Subjects respond to these statements using the same 7-point "Agree-Disagree" format previously described. For each subject, you sum responses to the five items to construct a single measure of intent to remain that can range from a low score of 5 to a high score of 35. With this measure, higher scores represent a higher level of intent to remain. When analyzing your data, use the SAS variable name "INT_REM" to represent this variable.

Data Set to be Analyzed

Table 10.E2.1 presents fictitious scores for each of the 22 subjects on each of the three variables in this study.

Table 10.E2.1

Variables Analyzed in the Nurses' Intent to Remain Study

Subject	Nurse assertiveness	Emotional exhaustion	Intent to remain
01	22	10	34
02	20	13	34
03	13	11	30
04	31	20	30
05	15	13	29
06	20	16	27
07	22	19	27
08	26	22	24
09	33	26	24
10	18	14	23
11	25	18	22
12	10	23	21
13	22	26	20
14	16	21	19
15	28	31	19
16	19	22	16
17	25	25	16
18	32	29	15
19	19	34	14
20	16	27	13
21	27	31	11
22	23	34	10

Table 10.E2.1 uses the standard format in which the rows represent the subjects, and the columns represent the variables. You can see that the first subject displays

- a Subject number of 01
- a score of 22 on Nurse assertiveness

- a score of 10 on Emotional exhaustion
- a score of 34 on Intent to remain.

Data for the remaining subjects can be interpreted in the same way.

Your Assignment Regarding the SAS Program

1. Write a SAS program that will input and analyze the data set presented in Table 10.E2.1. When you write the DATA step of this program, do the following:
   - Use the SAS programs presented in Chapter 10 of the *Student Guide* as models.
   - Type all of the information from Table 10.E2.1. You should type the subject number variable as the first column as it appears in the table. Name this variable SUB_NUM. When you type this SAS variable name, make sure you use an underscore (_) and not a hyphen (-). When you type subject numbers such as "01," make sure you type a zero (0) and not the letter O.
   - In the DATA step, remember to use the SAS variable names N_ASSERT, EXHAUST, and INT_REM to represent the three data variables.

2. When you write the PROC step of this program, do the following:
   - Use PROC PLOT to create a scattergram in which the criterion variable is intent to remain, and the predictor variable is nurse assertiveness.
   - Type your full name in the TITLE1 statement following this PROC step, so that it will appear in the output.
   - Use PROC CORR to compute all possible correlations between all three data variables: N_ASSERT, EXHAUST, and INT_REM. Do not include the variable SUB_NUM in this correlation matrix. This should be done using a single PROC CORR statement. For more information, see the section in Chapter 10 titled "Using PROC CORR to Compute All Possible Correlations for a Group of Variables."

3. Submit the program for analysis, and, if necessary, correct it so that it runs without errors.

4. Assume that you began with the following research question: "The purpose of your study was to determine whether nurse assertiveness is correlated with a nurse's intent to remain with the current employer." Your research hypothesis is that there will be a positive relationship between nurse assertiveness and the intent to remain. One of the correlations produced by the PROC CORR statement in your program is relevant to answering this question.

   Prepare a report summarizing the results of your analysis using the format shown in the section titled "Summarizing the Results of the Analysis" from Chapter 10 of the *Student Guide*. You will need to modify the report so that it is relevant to the analysis that you are conducting in this exercise. Underline each heading in this report, and do not forget to include information for sections A-L.

   Although your PROC CORR statement produced a number of correlations, you are to focus on only one for this exercise—the correlation between the following:

   (a) nurse assertiveness

   (b) the intent to remain.

What You Will Hand In

Hand in the following materials stapled together in this order:

1. A printout of your SAS program, your SAS log, and your SAS output files.

Your SAS output should include both

(a) the scattergram in which the criterion variable was intent to remain and the predictor variable was nurse assertiveness (you do not have to draw a "best fitting line" on this scattergram)

(b) the results of the PROC CORR statement that computed all possible correlations between the three data variables.

2. Your report summarizing the results of the analysis, including information for sections A-L from Chapter 10 of the *Student Guide*.

Hint

If your SAS program ran correctly, your output should resemble the following output:

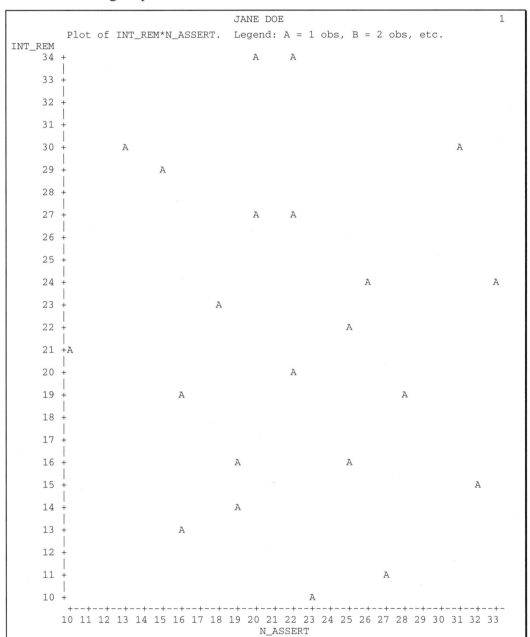

```
                                    JANE DOE                                   1
         Plot of INT_REM*N_ASSERT.  Legend: A = 1 obs, B = 2 obs, etc.
INT_REM
     34 +                              A     A
        |
     33 +
        |
     32 +
        |
     31 +
        |
     30 +         A                                                     A
        |
     29 +               A
        |
     28 +
        |
     27 +                              A     A
        |
     26 +
        |
     25 +
        |
     24 +                                                A                    A
        |
     23 +                        A
        |
     22 +                                             A
        |
     21 +A
        |
     20 +                                    A
        |
     19 +                  A                                   A
        |
     18 +
        |
     17 +
        |
     16 +                           A                 A
        |
     15 +                                                                  A
        |
     14 +                           A
        |
     13 +                  A
        |
     12 +
        |
     11 +                                                   A
        |
     10 +                                       A
         +--+--+--+--+--+--+--+--+--+--+--+--+--+--+--+--+--+--+--+--+--+--+--+-
        10 11 12 13 14 15 16 17 18 19 20 21 22 23 24 25 26 27 28 29 30 31 32 33
                                        N_ASSERT
```

```
                              JANE DOE                              2

                         The CORR Procedure

         3  Variables:     INT_REM  N_ASSERT EXHAUST

                          Simple Statistics

Variable    N        Mean    Std Dev          Sum     Minimum    Maximum

INT_REM    22   21.72727    7.15929    478.00000    10.00000   34.00000
N_ASSERT   22   21.90909    6.11718    482.00000    10.00000   33.00000
EXHAUST    22   22.04545    7.31259    485.00000    10.00000   34.00000

            Pearson Correlation Coefficients, N = 22
                    Prob > |r| under H0: Rho=0

                       INT_REM        N_ASSERT          EXHAUST

         INT_REM       1.00000        -0.11367         -0.86203
                                        0.6145           <.0001

         N_ASSERT     -0.11367         1.00000          0.37907
                        0.6145                           0.0819

         EXHAUST      -0.86203         0.37907          1.00000
                        <.0001          0.0819
```

Reference

Ellis, B. H., & Miller, K. I. (1993). The role of assertiveness, personal control, and participation in the prediction of nurse burnout. *Journal of Applied Communication Research, 21,* 327-342.

Exercises for Chapter 11: Bivariate Regression

Exercise 11.1: Predicting Current Drinking and Driving Behavior from Previous Behavior

Overview

Is it possible to predict a person's current drinking and driving behavior from previous incidents of drinking and driving? What percent of the variance in current behavior is accounted for by previous behavior? The current study was designed to answer these questions.

This exercise deals with the same study on drinking and driving that was described in Exercise 10.1, "Correlational Study of Drinking and Driving Behavior." In completing this exercise, you will analyze the same data set that you analyzed in Exercise 10.1 to determine whether the regression coefficient representing the relationship between drinking and driving behavior at Time 1 and drinking and driving behavior at Time 2 is significantly different from zero. You will use PROC PLOT to create a scattergram for your variables, and you will use PROC REG to perform a regression analysis on them. In addition, you will prepare a report that summarizes the results of your analysis.

If you have already completed Exercise 10.1 (and if you saved the data set to a disk or other media), then it will not be necessary to type the data set again; you may simply perform new analyses on the same data. If you have not already completed Exercise 10.1, see the following section for a description of the study.

Note: Although the investigation reported here is fictitious, it was inspired by the actual study reported by Green (1991).

The Study

Suppose that you conduct a correlational study on a group of 21 adults (subjects) who sometimes drink alcoholic beverages. Your study takes place during the months of June and July one summer, and you obtain data at two points in time.

Your study includes three predictor variables: moral commitment to the legal norm, perceived certainty of arrest, and drinking and driving behavior at the end of June (Time 1). The single criterion variable in your study is drinking and driving behavior at the end of July (Time 2). From your sample study, you obtain scores for 21 subjects based on these variables as follows:

Moral commitment to the legal norm. The legal norm holds that driving a car while drinking alcohol is wrong. Therefore, people who score high on moral commitment to the legal norm will tend to agree that drinking and driving is wrong.

Suppose that you measure "moral commitment" with a 4-item scale. The scale consists of statements such as "It is wrong to drive a car while under the influence of alcohol." Subjects respond to these statements using a 7-point scale in which "1" represents "Disagree Very Strongly" and "7" represents "Agree Very Strongly." For each subject, you sum responses to the four items to construct a single measure of moral commitment that can range from a low score of 4 to a high score of 28. With this measure, higher scores represent a higher level of commitment to the legal norm.

In analyzing your data, you use the SAS variable name "MC_T1" to represent this variable. Here, MC_T1 stands for "moral commitment at Time 1."

Perceived certainty of arrest. Assume that you also use a 4-item scale to measure the subjects' belief that they will be arrested if they drink and drive.

The scale consists of statements such as "There is a good chance I will be arrested if I drive a car under the influence of alcohol." Again, they respond to these statements using a 7-point scale in which "1" represents "Disagree Very Strongly" and "7" represents "Agree Very Strongly." You sum responses to the four items to construct a single measure of perceived certainty. This variable can range from a low score of 4 to a high score of 28. With this measure, higher scores represent greater certainty that they will be arrested if they drink and drive.

In analyzing your data, you use the SAS variable name "PC_T1" to represent this variable. Here, PC_T1 stands for "perceived certainty at Time 1."

Drinking and driving at Time 1. At the end of the month of June, you administer the previously mentioned scales that assess moral commitment to the legal norm and perceived certainty of arrest. You also ask subjects to indicate the number of times that they drank and drove during the month of June. When you analyze your data, you will determine whether this variable is correlated with the number of times that they drank and drove during the following month (July).

In analyzing your data, you use the SAS variable name "DD_T1" to represent this variable. Here, DD_T1 stands for "drinking and driving at Time 1."

Drinking and driving at Time 2. At the end of the following month (July), you ask subjects to again indicate the number of times that they drank and drove during that month. This measure will serve as the primary criterion variable that you are trying to predict.

In analyzing your data, you use the SAS variable name "DD_T2" to represent this variable. Here, DD_T2 stands for "drinking and driving at Time 2."

Data Set to be Analyzed

In this exercise, you will analyze the same data set that appeared in Table 10.E1.1 from Exercise 10.1 in this book. If you have already typed that data (and have it saved to a diskette or other media), you may simply re-analyze the same data set using your original DATA step. You will delete the PROC statements that were unique to Exercise 10.1, and add new PROC statements for the current exercise as described in the following section. If you have not already typed the data set from Exercise 10.1, you should refer to Exercise 10.1 for instructions on typing the data set before continuing this exercise.

Your Assignment Regarding the SAS Program

1. Follow the directions provided in the previous section titled "Data Set to be Analyzed" to write the DATA step for this exercise.
2. In the PROC step of your program, add statements that will perform the following tasks (all of these tasks should be done within a single program):
   - Use PROC PLOT to create a scattergram in which the criterion variable is drinking and driving behavior at Time 2, and the predictor variable is drinking and driving behavior at Time 1.
   - Type your full name in the TITLE1 statement following this PROC, so that it will appear in the output.
   - Use PROC REG to perform a regression analysis in which the criterion variable is drinking and driving behavior at Time 2, and the predictor variable is drinking and driving behavior at Time 1. In the MODEL statement, be sure to include the keywords "STB" and "P," as was done in the example programs in Chapter 11 of the *Student Guide*.

   In writing this program, do the following:

   - Use the SAS programs presented in Chapter 11 of the *Student Guide* as models.
   - In the DATA step, remember to use the SAS variable names MC_T1, PC_T1, DD_T1, and DD_T2 to represent the four data variables.
3. Submit the program for analysis, and, if necessary, correct it so that it runs without errors.

Your Assignment Regarding the SAS Output

1. Draw a best-fitting regression line through the scattergram that was generated by the PROC PLOT statement in your SAS program. To do this, you will follow the directions that are provided in Chapter 11 of the

Student Guide in the section titled "Drawing a Regression Line Through the Scattergram." (Be sure to read those directions carefully as you complete the following tasks.) Specifically, do the following:

- At the top of a clean sheet of paper write the title, "Computing a predicted value of Y that is associated with a low value of X." Below this title you will write the regression equation from the analysis that you have just performed. This regression equation will take this form:

  $$Y' = b\,(X) + a$$

  You will fill in the appropriate values for b and a in the preceding equation. You will obtain these values from the results of the PROC REG that you have just performed.
- Choose a relatively low value of X and insert it in the preceding equation. Then solve for Y′. In solving for Y′, you should show at least two steps of work.
- On the same sheet of paper, write the title "Computing a predicted value of Y that is associated with a high value of X." Below this title, again write the regression equation from the analysis that you have just performed.
- Choose a relatively high value of X and insert it in the preceding equation. Then solve for Y′. In solving for Y′, you should show at least two steps of work.
- Now, take the page of SAS output that contains the scattergram generated by your PROC PLOT statement. Draw a best-fitting regression line through this scattergram. Do this by following the directions provided in Chapter 11 of the *Student Guide* in the section titled "Drawing a Regression Line Through the Scattergram." For example, you will place a dot or a small "x" on your scattergram that corresponds to the low Y′ value that you previously computed. In doing this, make your dot or small "x" large enough so that the person grading this exercise will be able to easily see it and verify that you placed it in the correct location. You will also place a dot or a small "x" on your scattergram that corresponds to the high Y′ value that you previously computed. You should also draw this dot or small "x" so that it will be easy to see and grade. Then draw a regression line that

connects these two dots according to the instructions provided in your *Student Guide*.

2. Assume that you begin with the following statement of a research question: "The purpose of this study was to determine whether the regression coefficient representing the relationship between drinking and driving behavior at Time 1 (during the month of June) and drinking and driving at Time 2 (during the month of July) is significantly different from zero." Your research hypothesis is that there will be a positive relationship between drinking and driving behavior at Time 1 and drinking and driving behavior at Time 2. The output generated by the PROC REG statement in your SAS program will be relevant to this research question.

   Prepare a report summarizing the analysis, using the format shown in the section titled "Summarizing the Results of the Analysis" from Chapter 11 of the *Student Guide*. You will have to modify the report from that section so that it is relevant to the analysis that you are actually conducting in this exercise. Underline each heading in this report, and remember to include information for sections A-L.

   Remember that you are investigating the relationship between

   - drinking and driving behavior at Time 1
   - drinking and driving behavior at Time 2.

What You Will Hand In

Hand in the following materials stapled together in this order:

1. A printout of your SAS program (including data), your SAS log, and your SAS output files. Your SAS output should include the following (arranged in this order):

   (a) The scattergram produced by PROC PLOT in which the criterion variable was drinking and driving behavior at Time 2 and the predictor variable was drinking and driving behavior at Time 1. You must draw a best-fitting regression line through this scattergram according to the directions provided in the previous section.

(b) The results of the PROC REG statement in which the criterion variable was drinking and driving behavior at Time 2 and the predictor variable was drinking and driving behavior at Time 1. This output should include both:
   - the standard output for PROC REG, similar to Output 11.1 from Chapter 11 (in other words, the "Analysis of Variance" section and the "Parameter Estimates" section)
   - the table of predicted values and residuals, similar to Output 11.9 from Chapter 11. Remember that this table is requested by including the keyword "P" in your MODEL statement.

2. Your report summarizing the results of the analysis. The example reports in the *Student Guide* include a copy of the scattergram produced by PROC PLOT with the regression line drawn in. However, because the materials that you hand in for step 1 in this section should already contain your scattergram with regression line, you do not need to include a second copy for this step as well.
3. A sheet of paper with a section titled "Computing a predicted value of Y′ that is associated with a low value of X" that shows the math that you performed when computing a value of Y′ that was associated with a low value of X. This sheet should also have a section titled "Computing a predicted value of Y′ that is associated with a high value of X" that shows the math that you performed when computing a value of Y′ that was associated with a high value of X.

Hint

If your SAS program ran correctly, your output should resemble the following excerpt of output. Remember that your program will produce more output than shown in this excerpt.

```
                              JOHN DOE                                2

                         The REG Procedure
                           Model: MODEL1
                    Dependent Variable: DD_T2

                        Analysis of Variance

                                  Sum of           Mean
Source                  DF       Squares         Square    F Value    Pr > F

Model                    1      36.44885       36.44885       9.12    0.0003
Error                   19      36.21782        1.90620
Corrected Total         20      72.66667


        Root MSE              1.38065    R-Square     0.5016
        Dependent Mean        4.66667    Adj R-Sq     0.4754
        Coeff Var            29.58541


                         Parameter Estimates

                  Parameter   Standard                        Standardized
Variable   DF      Estimate      Error   t Value   Pr > |t|       Estimate

Intercept   1       1.47529    0.78957       .87     0.0772              0
DD_T1       1       0.70546    0.16133      4.37     0.0003        0.70823
```

Reference

Green, D. E. (1991). Inhibition, motivation, and self-reported involvement in drinking and driving behavior. *Criminal Justice Review, 16,* 1-16.

Exercise 11.2: Predicting Nurses' Intent to Remain from Emotional Exhaustion

Overview

In Exercise 10.2, you computed the correlation between (a) nurses' intent to remain with their current employer and (b) nurses' assertiveness. In this exercise, you will analyze the same data set, but will focus on a different predictor variable. Here, the criterion variable will again be the intent to remain, but the predictor variable this time will be a measure of the nurses' emotional exhaustion.

You will use PROC PLOT to create a scattergram for the two variables, and you will use PROC REG to regress intent to remain on emotional exhaustion. You will review the results of PROC REG to test the null hypothesis that the regression coefficient is equal to zero in the population, and you will prepare an analysis report to summarize your findings. Finally, you will draw a regression line through the scattergram created by the PLOT procedure.

In completing this exercise, you will analyze the same data set that you analyzed in Exercise 10.2. If you have already completed Exercise 10.2 (and if you saved the data set to a disk or other media), then it will not be necessary to type the data set again; you may simply perform new analyses on the same data. If you have not already completed Exercise 10.2, see the following section for a description of the study.

Note: Although the study reported here is fictitious, it was inspired by the actual study reported by Ellis and Miller (1993).

The Study

Suppose that you conduct a study to investigate variables that may predict nurses' intent to remain with their current employers. You develop a questionnaire that assesses the nurses' intent to remain, as well as their assertiveness and emotional exhaustion.

You administer the questionnaire to 22 nurses (subjects) who are currently employed in hospitals and similar settings. You type the data and analyze it using PROC PLOT and PROC REG to better understand the relationship between the criterion variable and the predictor variables. Your research hypothesis is that there will be a negative relationship between nurses' emotional exhaustion and their intent to remain.

Your study includes two predictor variables: nurse assertiveness and emotional exhaustion. The single criterion variable is the nurses' intent to remain with their current employer. From your sample study, you obtain scores for 22 subjects based on these variables as follows:

Nurse assertiveness. Nurse assertiveness refers to the extent to which nurses speak up, voice their opinions, express their sense of self-worth, and validate their own professional status (particularly in interactions with physicians). Assume that you assess nurse assertiveness using a 5-item scale. The scale consists of statements such as "I tell physicians when, in my opinion, the approach they are using to treat a patient is not likely to be effective." Subjects respond to these statements using a 7-point scale in which "1" represents "Disagree Very Strongly" and "7" represents "Agree Very Strongly." For each subject, you sum responses to the five items to construct a single measure of nurse assertiveness that can range from a low score of 5 to a high score of 35. With this measure, higher scores represent a higher level of nurse assertiveness. In analyzing your data, you use the SAS variable name "N_ASSERT" to represent this variable.

Emotional exhaustion. Emotional exhaustion refers to the extent to which nurses are psychologically fatigued from the stress and other problems experienced in their jobs. Assume that you assess emotional exhaustion using a 5-item scale. The scale consists of statements such as "I feel tired when I think about having to face another day at work." Subjects respond to these statements using the same 7-point "Agree-Disagree" format previously described. For each subject, you sum responses to the five items to construct a single measure of emotional exhaustion that can range from a low score of 5 to a high score of 35. With this measure, higher scores represent a higher level of emotional exhaustion. In analyzing your data, you use the SAS variable name "EXHAUST" to represent this variable.

Intent to remain. Intent to remain refers to the extent to which nurses plan to remain employed with their current organizations. Assume that you assess intent to remain using a 5-item scale. The scale consists of statements such as "I plan to remain employed at this hospital for a long time." Subjects respond to these statements using the same 7-point "Agree-Disagree" format previously described. For each subject, you sum responses to the five items to construct a single measure of intent to remain that can range from a low score of 5 to a high score of 35. With this measure, higher scores represent a higher level of intent to remain. In analyzing your data, you use the SAS variable name "INT_REM" to represent this variable.

Data Set to be Analyzed

In this exercise, you will analyze the same nurses' intent to remain data set that appeared in Table 10.E2.1 from Exercise 10.2 in this book. If you have already typed that data (and have it saved to a disk or other media), you may simply re-analyze the same data set using your original DATA step. You will delete the PROC statements that were unique to Exercise 10.2, and add new PROC statements for the current exercise as described in the following section. If you have not already typed the data set from Exercise 10.2, you should refer to Exercise 10.2 for instructions on typing the data set before continuing this exercise.

Your Assignment Regarding the SAS Program

1. Follow the directions provided in the previous section titled "Data Set to be Analyzed" to write the DATA step for this exercise.

2. In the PROC step of your program, add statements that will perform the following tasks (all of these tasks should be done within a single program):
   - Use PROC PLOT to create a scattergram in which the criterion variable is intent to remain, and the predictor variable is emotional exhaustion.
   - Type your full name in the TITLE1 statement following this PROC, so that it will appear in the output.
   - Use PROC REG to perform a regression analysis in which the criterion variable is intent to remain, and the predictor variable is emotional exhaustion. In the MODEL statement, be sure to include the keywords "STB" and "P," as was done in the example programs in Chapter 11 of the *Student Guide*.

   In writing this program, do the following:
   - Use the SAS programs presented in Chapter 11 of the *Student Guide* as models.
   - In the DATA step, remember to use the SAS variable names SUB_NUM, N_ASSERT, EXHAUST, and INT_REM to represent the subject number variable and the three data variables.
3. Submit the program for analysis, and, if necessary, correct it so that it runs without errors.

Your Assignment Regarding the SAS Output

1. Draw a best-fitting regression line through the scattergram that was generated by the PROC PLOT statement in your SAS program. To do this, you will follow the directions that are provided in Chapter 11 of the *Student Guide* in the section titled "Drawing a Regression Line Through the Scattergram." (Be sure to read those directions carefully as you complete the following tasks.) Specifically, do the following:
   - At the top of a clean sheet of paper write the title, "Computing a predicted value of Y that is associated with a low value of X." Below this title you will write the regression equation from the

analysis that you have just performed. This regression equation will take this form:

$$Y' = b\,(X) + a$$

You will fill in the appropriate values for *b* and *a* in the preceding equation. You will obtain these values from the results of the PROC REG that you have just performed.

- Choose a relatively low value of X and insert it in the preceding equation. Then solve for Y′. In solving for Y′, you should show at least two steps of work.
- On the same sheet of paper, write the title "Computing a predicted value of Y that is associated with a high value of X." Below this title, again write the regression equation from the analysis that you have just performed.
- Choose a relatively high value of X and insert it in the preceding equation. Then solve for Y′. In solving for Y′, you should show at least two steps of work.
- Now, take the page of SAS output that contains the scattergram generated by your PROC PLOT statement. Draw a best-fitting regression line through this scattergram. Do this by following the directions provided in Chapter 11 of the *Student Guide* in the section titled "Drawing a Regression Line Through the Scattergram." For example, you will place a dot or a small "x" on your scattergram that corresponds to the low Y′ value that you previously computed. In doing this, make your dot or small "x" large enough so that the person grading this exercise will be able to easily see it and verify that you placed it in the correct location. You will also place a dot or a small "x" on your scattergram that corresponds to the high Y′ value that you previously computed. You should also draw this dot or small "x" so that it will be easy to see and grade. Then draw a regression line that connects these two dots according to the instructions provided in your *Student Guide*.

2. Assume that you begin with the following statement of a research question: "The purpose of this study was to determine whether the regression coefficient representing the relationship between nurses' emotional exhaustion and their intent to remain is significantly different

from zero." The output generated by the PROC REG statement in your SAS program will be relevant to this research question.

Prepare a report summarizing the analysis, using the format shown in the section titled "Summarizing the Results of the Analysis" from Chapter 11 of the *Student Guide*. You will have to modify the report from that section so that it is relevant to the analysis that you are actually conducting in this exercise. Underline each heading in this report, and remember to include information for sections A-L.

Remember that you are investigating the relationship between

- emotional exhaustion
- intent to remain.

What You Will Hand In

Hand in the following materials stapled together in this order:

1. A printout of your SAS program (including data), your SAS log, and your SAS output files. Your SAS output should include the following (arranged in this order):
   (a) The scattergram in which the criterion variable was intent to remain and the predictor variable was emotional exhaustion. You must draw a best-fitting regression line through this scattergram according to the directions provided in the previous section.
   (b) The results of the PROC REG statement in which the criterion variable was intent to remain and the predictor variable was emotional exhaustion. This output should include both:
      - the standard output for PROC REG, similar to Output 11.1 from Chapter 11 (in other words, the "Analysis of Variance" section and the "Parameter Estimates" section)
      - the table of predicted values and residuals, similar to Output 11.9 from Chapter 11. Remember that this table is requested by including the keyword "P" in your MODEL statement.
2. Your report summarizing the results of the analysis. The example reports in the *Student Guide* include a copy of the scattergram produced by PROC

PLOT with the regression line drawn in. However, because the materials that you hand in for step 1 in this section should already contain your scattergram with regression line, you do not need to include it in this step as well.

3. A sheet of paper titled "Computing a predicted value of Y that is associated with a low value of X." This sheet should show the math that you performed when computing a value of Y′ that was associated with a low value of X, as well as a value of Y′ that was associated with a high value of X.

Hint

If your SAS program ran correctly, your output should resemble the following excerpt of output. Remember that your program will produce more output than shown in this excerpt.

```
                              JANE DOE                                     2

                          The REG Procedure
                            Model: MODEL1
                    Dependent Variable: INT_REM

                         Analysis of Variance

                                   Sum of           Mean
Source                  DF        Squares         Square    F Value    Pr > F

Model                    1      799.84269      799.84269      57.85    <.0001
Error                   20      276.52095       13.82605
Corrected Total         21     1076.36364


           Root MSE              3.71834    R-Square     0.7431
           Dependent Mean       21.72727    Adj R-Sq     0.7303
           Coeff Var            17.11370

                         Parameter Estimates

                 Parameter   Standard                        Standardized
Variable   DF     Estimate      Error   t Value   Pr > |t|       Estimate

Intercept   1     40.33273    2.57142     15.68     <.0001              0
EXHAUST     1     -0.84396    0.11096     -7.61     <.0001       -0.86203
```

Reference

Ellis, B. H., & Miller, K. I. (1993). The role of assertiveness, personal control, and participation in the prediction of nurse burnout. *Journal of Applied Communication Research, 21,* 327-342.

EXERCISES 12

Exercises for Chapter 12: Single-Sample *t* Test

Exercise 12.1: Answering SAT Reading Comprehension Questions Without the Passages

Overview

In this exercise, you will read about a fictitious study in which subjects respond to test items from the Reading Comprehension section of the Scholastic Aptitude Test without first reading the passages on which the test items are based. You will be provided with fictitious data from this study, and you will prepare a SAS program that inputs this data. You will perform a single-sample *t* test to determine whether the mean score demonstrated by your sample of subjects is significantly different from the population mean stated under the null hypothesis. Finally, you will prepare an analysis report that summarizes your findings.

The Study

The Scholastic Aptitude Test (SAT) is a standardized test that is widely used to assess an individual's ability to do college-level work. The Reading Comprehension section of the SAT is a subset of items used to assess the individual's ability to read and understand short passages of text. This task involves presenting the individual with several paragraphs of prose, followed by a number of multiple-choice items that assess the individual's understanding of the material that he or she has just read.

But what would happen if you asked individuals to respond to the reading comprehension items without first allowing them to read the relevant passages of text? At first glance, you might expect the individuals to perform at chance level. That is, if each multiple choice question provides 5 possible responses, you might expect the individuals' average score to be only 20 percent correct. This is because they would be selecting their answers randomly, and when individuals select answers randomly on a test that uses 5 alternatives for each question, they tend to be correct only 1 time out of 5 on the average, or 20 percent of the time.

It is possible, however, that the average individual will score at above-chance levels. Some critics have argued that scores on the SAT Reading Comprehension section are influenced by factors other than the individuals' reading comprehension skills only. For example, some people might obtain good scores on this task simply because they have strong verbal skills and can identify the correct answer in a multiple-choice question even when they have not read the material that the question deals with. If this is the case, then these individuals should score at higher-than-chance levels on SAT reading comprehension items, even if they are not given the passages of text that go with the items.

Research method. To see if this is possible, suppose that you conduct the following study: You ask 17 subjects to each respond to 100 multiple choice items taken from the Reading Comprehension section of the SAT. They follow essentially the same test-taking procedures normally followed when the SAT is administered, with one important exception: they are not allowed to read the passages of text that the comprehension items are based on.

Possible results. If these items are pure measures of reading comprehension, then the subjects should respond to them in a random, haphazard manner, and should get approximately 20 percent of them correct, only due to chance (again, this is because each item has 5 possible answers, and the subject has a 1 in 5 chance of getting a given item correct). Because the test has 100 items, the average subject is therefore expected to get only 20 items correct.

However, if the average score for this sample is significantly higher than 20 items correct, it might mean that these reading comprehension items are measuring a skill in addition to reading comprehension. To find out if this is the case, you will have your 17 subjects respond to the items, obtain their average score (out of 100), and use a single-sample *t* test to determine whether this average score is significantly different from 20.

Note: Although the study described here is fictitious and the results are exaggerated, the research design was inspired by an actual study reported by Katz et al. (1990).

Data Set to be Analyzed

Table 12.E1.1 presents the (fictitious) scores on the criterion variable: the number of items that each subject answered correctly out of a possible 100 (there were 17 subjects, therefore there are 17 scores listed in the table).

Table 12.E1.1

Number of Items that Subjects Answered Correctly Out of 100

Subject	Score
01	49
02	46
03	51
04	53
05	55
06	50
07	43
08	53
09	49
10	51
11	49
12	47
13	57
14	51
15	47
16	45
17	45

Your Assignment

1. Create a SAS program that will input the data set from Table 12.E1.1 and analyze it using PROC TTEST to determine whether the average score for your 17 subjects was significantly different from 20. When you write this program, do the following:
   - Use the SAS programs presented in Chapter 12 of the *Student Guide* as models.
   - Use the SAS variable name SUB_NUM to represent subject numbers, and the SAS variable name SCORE to represent your raw criterion variable (the number correct out of 100).
   - Determine the "comparison number" for your PROC TTEST statement using the information provided in this exercise.
   - Also, in the PROC TTEST statement, write the ALPHA option so that it will request the 95 percent confidence interval for the mean (this was done with all of the programs in Chapter 12 of the *Student Guide*).
   - Type your full name in the TITLE1 statement, so that it will appear in the output.
2. Submit the program for analysis, and, if necessary, correct it so that it runs without errors.
3. At the top of a clean sheet of paper, write the title "Computing the Index of Effect Size." On this sheet of paper, compute the index of effect size (*d*) for the current analysis. You will have to do this by hand; see Chapter 12 of the *Student Guide* for details about computing the index of effect size.

   In performing this task, first provide the symbol for effect size (*d*), and then the formula for effect size that was provided in Chapter 12 of the *Student Guide*. Following that, copy the formula again, this time substituting the actual values from the SAS output for the symbols. For example, this means that you should substitute the actual value of the estimated population standard deviation (3.7603) for the symbol (s_x). If the values that you are inserting in the formula have decimal places, use the same number of decimal places that these values actually had in the SAS output. Next, complete the steps that are necessary to solve for *d* by performing only one mathematical operation at each step. Your last step should provide your final value of *d*, rounded to two decimal places only.

4. Assume that you began with the following research question: "The purpose of this study was to determine whether subjects answering SAT reading comprehension items without passages would perform at a level that is higher than the level expected from random responses." The output created by your SAS program should be relevant to this research question.

   On a separate sheet of paper, prepare a report summarizing the analysis, using the format shown in the section "Summarizing the Results of the Analysis" from Chapter 12 of the *Student Guide*. You will have to modify the report so that it is relevant to the present analysis. Underline each heading in this report, and remember to include information for sections A–M.

What You Will Hand In

Hand in the following materials stapled together in this order:

1. A printout of your SAS program (including data), your SAS log, and your SAS output files.
2. A page titled "Computing the Index of Effect Size." This page should contain everything requested by Step 3 in the section titled "Your Assignment" of this exercise.
3. A separate page containing your report summarizing the results of the analysis, including information for sections A–M.

Hint

If your SAS program ran correctly, your output should resemble the following output:

```
                              JANE DOE                                  1
                         The TTEST Procedure

                              Statistics

              Lower CL              Upper CL Lower CL           Upper CL
Variable   N      Mean       Mean       Mean  Std Dev   Std Dev  Std Dev

SCORE     17    47.537     49.471     51.404   2.8005    3.7603   5.7229

                              Statistics

              Variable   Std Err     Minimum     Maximum

              SCORE        0.912          43          57

                               T-Tests

               Variable        DF     t Value     Pr > |t|

               SCORE           16       32.31       <.0001
```

Reference

Katz, S., Lautenschlager, A., Blackburn, A.B., & Harris, F.H. (1990). Answering reading comprehension items without passages on the SAT. *Psychological Science, 1,* 122-127.

Exercise 12.2: Predicting the Results of Coin Flips

Overview

In this exercise, you will read about a fictitious study in which subjects attempt to predict the outcome of 20 individual coin flips. You will be provided with fictitious data from this study, and you will prepare a SAS program that inputs this data. You will perform a single-sample *t* test to determine whether the mean number of correct predictions demonstrated by your subjects is significantly different from the number that would be expected with random guessing. Finally, you will prepare an analysis report that summarizes your findings.

The Study

This exercise is a modification of the fictitious ESP study described at the beginning of Chapter 12 in the *Student Guide*. There, you were told to suppose that you were conducting research with the ESP Club on campus. For the current exercise, you should suppose that the club includes 14 members who claim to be able to predict the future by means of precognition. To prove it, they each complete 20 trials in which they predict the results of a coin flip. This means that, for each subject, a coin is flipped 20 times and the subject predicts whether the coin will come up heads or tails prior to each flip. If the results of each coin flip match what the subject predicted, then the subject has scored a "correct" prediction for that flip.

The club members show some variability in their performance. One member achieves a score of only 7 correct predictions, another achieves a score of 15 correct predictions, and so forth. When you average their scores, you find that the average for the group is 10.857. This means that, on the average, the 14 members guessed correctly on 10.857 out of 20 coin flips. Members of the ESP Club are happy with this average score. They point out that, if they did not have precognition skills, the probability of correctly guessing a coin flip was only .5. This means that, out of 20 flips, they should have made an average of only 10 correct guesses. Their obtained average of 10.857 correct predictions is slightly above the .5 probability score.

It is true that their sample mean of 10.857 correct guesses is higher than the hypothetical population mean of 10 correct guesses, but is it significantly higher? To find out, you perform a single-sample *t* test, testing the null hypothesis that the sample mean came from a population in which the population mean was actually equal to 10. If you reject this null hypothesis, it will provide some support for the club members' claim that they have ESP.

Data Set to be Analyzed

Table 12.E2.1 presents the (fictitious) scores on the criterion variable: the number of correct predictions that each subject made out of 20 coin flips (there were 14 subjects, therefore there are 14 scores listed in the table).

Table 12.E2.1

Number of Correct Predictions Out of 20 Coin Flips

Subject	Correct
01	11
02	12
03	9
04	9
05	10
06	8
07	14
08	13
09	10
10	7
11	11
12	12
13	11
14	15

Your Assignment

1. Create a SAS program that will input the data set from Table 12.E2.1 and analyze it using PROC TTEST to determine whether the average score for your 14 subjects was significantly different from 10. When you write this program, do the following:
   - Use the SAS programs presented in Chapter 12 of the *Student Guide* as models.
   - Use the SAS variable name SUB_NUM to represent subject numbers, and the SAS variable name CORRECT as the name for your raw criterion variable (the number of correct predictions out of 20 coin flips).
   - Determine the "comparison number" for your PROC TTEST statement using the information provided in this exercise.
   - Also, in the PROC TTEST statement, write the ALPHA option so that it will request the 95 percent confidence interval for the mean (this was done with all of the programs in Chapter 12 of the *Student Guide*).
   - Type your full name in the TITLE1 statement, so that it will appear in the output.

2. Submit the program for analysis, and, if necessary, correct it so that it runs without errors.

3. At the top of a clean sheet of paper, write the title "Computing the Index of Effect Size." On this sheet of paper, compute the index of effect size (*d*) for the current analysis. You will have to do this by hand; see Chapter 12 of the *Student Guide* for details about computing the index of effect size.

   In performing this task, first provide the symbol for effect size (*d*), and then the formula for effect size that was provided in Chapter 12 of the *Student Guide*. Following that, copy the formula again, this time substituting the actual values from the SAS output for the symbols. For example, this means that you should substitute the actual value of the estimated population standard deviation for its symbol (s_x). If the values that you are inserting in the formula have decimal places, use the same number of decimal places that these values actually had in the SAS output. Next, complete the steps that are necessary to solve for *d* by performing only one mathematical operation at each step. Your last step should provide your final value of *d*, rounded to two decimal places only.

4. Assume that you began with the following research question: “The purpose of this study was to determine whether subjects predicting the results of coin flips would perform at a level that is higher than the level expected from random guesses.” The output created by your SAS program should be relevant to this research question.

   On a separate sheet of paper, prepare a report summarizing the analysis, using the format shown in the section “Summarizing the Results of the Analysis” from Chapter 12 of the *Student Guide*. You will have to modify the report so that it is relevant to the present analysis. Underline each heading in this report, and remember to include information for sections A–M from Chapter 12.

What You Will Hand In

Hand in the following materials stapled together in this order:

1. A printout of your SAS program (including data), your SAS log, and your SAS output files.
2. A page titled “Computing the Index of Effect Size.” This page should contain everything requested by Step 3 in the section titled “Your Assignment” of this exercise.
3. A separate page containing your report summarizing the results of the analysis, including information for sections A–M from Chapter 12.

Hint

If your SAS program ran correctly, your output should resemble the following output:

```
                              JANE DOE                                     1

                       The TTEST Procedure

                              Statistics

             Lower CL           Upper CL  Lower CL           Upper CL
Variable   N     Mean      Mean      Mean   Std Dev  Std Dev  Std Dev

CORRECT   14    9.559    10.857    12.155    1.6299   2.2483   3.6221

                              Statistics

                Variable   Std Err    Minimum    Maximum

                CORRECT     0.6009          7         15


                              T-Tests

               Variable       DF    t Value     Pr > |t|

               CORRECT        13       1.43       0.1773
```

Exercises for Chapter 13: Independent-Samples *t* Test

Exercise 13.1: Sex Differences in Sexual Jealousy

Overview

In this exercise, you will read about a fictitious study in which male and female subjects are asked to imagine how they would feel if their romantic partners had intercourse with another person. Subjects then rate how distressed they would feel in this situation. You will be provided with fictitious data from this study, and you will prepare a SAS program that inputs this data. You will perform an independent-samples *t* test to determine whether there is a significant difference between male subjects and female subjects with respect to their distress scores. Finally, you will prepare an analysis report that summarizes your findings.

The Study

Some researchers in the field of sociobiology have hypothesized that there might be differences between men and women with respect to jealousy. Specifically, some have argued that there might be a difference between men and women with respect to the types of jealousy-provoking situations that cause them the greatest distress.

For example, some have predicted that women should display higher levels of **romantic jealousy** than men. This means that, if you ask a group of subjects to imagine that their partners have developed a deep emotional attachment to someone else, the female subjects in this group should be more distressed by this thought than the male subjects. By the same token, it has also been predicted that men should display higher levels of **sexual jealousy** than women. This means that, if you ask a group of subjects to imagine that their partners have had sexual intercourse with someone else, the male subjects in this group should be more distressed by this thought than the female subjects.

Suppose that you conduct a study to investigate the second of these two predictions (i.e., suppose that you develop a study to investigate *sexual* jealousy). Specifically, suppose that you conduct a study in which you ask both male and female subjects to imagine how distressed they would feel if their partners had sexual intercourse with another person. The subjects then rate their level of psychological distress. You then analyze the resulting data to determine whether there is a significant difference between male subjects and female subjects with respect to their scores on this measure of psychological distress.

Note: Although the study described here is fictitious, it was inspired by the actual study reported by Buunk, et al. 1996.

Research question. The purpose of this study was to ask a group of subjects to imagine how they would feel if they learned that their partners had had intercourse with someone else, and then determine whether male subjects would score higher than female subjects on a measure of psychological distress.

Research hypothesis. When asked to imagine how they would feel if they learned that their partner had had sexual intercourse with someone else, male subjects will display higher mean scores than female subjects on a measure of psychological distress.

Research Method. In your study, subjects individually read written instructions that tell them to imagine how they would feel if they learned that their partner had had intercourse with someone else. After imagining this, they rate how distressed they would feel by responding to the following 4 items (for each item, they would circle a number between 1 and 7 to indicate how they would feel):

```
Not at all distressed  1  2  3  4  5  6  7  Extremely distressed

      Not at all upset  1  2  3  4  5  6  7  Extremely upset

     Not at all angry  1  2  3  4  5  6  7  Extremely angry

      Not at all hurt  1  2  3  4  5  6  7  Extremely hurt
```

For a given subject, you could sum his or her responses to the four items to create a single "distress" score. Scores for this measure of distress could range from a low of 4 (if the subject circled "1s" for each item) to a high of 28 (if the subject circled "7s" for each item). For this study, higher scores would indicate higher levels of psychological distress.

Predictor and criterion variables in the analysis. The predictor variable in your study is "subject sex." This is a nominal-level variable that consists of just two values: female and male. In your analysis, you will give this variable the SAS variable name "SEX," and will use the symbol "F" to represent female subjects and "M" to represent male subjects.

The criterion variable in your study is "distress." This is the subjects' scores on the 4-item measure of psychological distress previously described. You can assume that this variable is on an interval scale, as it has approximately equal intervals, but no true zero point. In the analysis, you will give it the SAS variable name "DISTRESS."

Data Set to be Analyzed

A random sample of 18 women and 18 men individually complete this task. Table 13.E1.1 presents their scores for the measure of distress.

Table 13.E1.1

Subject Sex and Scores for the Measure of Distress

Subject	Subject sex[a]	Distress
01	M	22
02	M	25
03	M	23
04	M	24
05	M	20
06	M	28
07	M	27
08	M	23
09	M	23
10	M	24
11	M	26
12	M	26
13	M	25
14	M	21
15	M	22
16	M	23
17	M	24
18	M	25
19	F	22
20	F	22
21	F	25
22	F	18
23	F	23
24	F	24
25	F	19
26	F	20
27	F	20
28	F	20
29	F	21
30	F	21
31	F	21
32	F	19
33	F	19
34	F	22
35	F	23
36	F	21

[a] With the variable "Subject sex," the value "M" represents male subjects and the value "F" represents female subjects.

In Table 13.E1.1, you can see that Subject 01 has the value "M" in the "Subject sex" column, indicating that this subject is a male. Subject 01 has the value "22" in the "Distress" column, indicating that his score for the measure of psychological distress was 22. If you look at Subject 19, you can see that this subject has the value "F" in the "Subject sex" column, indicating that she is a female. Subject 19 has the value "22" in the "Distress" column, which means that her score for the measure of psychological distress was also 22.

Your Assignment

1. Create a SAS program that will input data from Table 13.E1.1 and analyze it using PROC TTEST to determine whether there is a significant difference between male and female subjects with respect to their mean scores on the distress variable. When you write this program, do the following:

   - Use the SAS programs presented in Chapter 13 of the *Student Guide* as models.
   - Use the SAS variable name SUB_NUM to represent subject numbers.
   - Use the SAS variable name SEX for subject sex. Use the value "M" to represent the male subjects and "F" to represent the female subjects (as in Table 13.E1.1).
   - Use the SAS variable name DISTRESS to represent subject scores for the distress variable.
   - Type your full name in the TITLE1 statement, so that it will appear in the output.

2. Submit the program for analysis, and, if necessary, correct it so that it runs without errors.

3. At the top of a clean sheet of paper, write the title "Computing the Index of Effect Size." On this sheet of paper, compute the index of effect size (*d*) for the current analysis. You will have to do this by hand; see Chapter 13 of the *Student Guide* for details about computing the index of effect size.

   In performing this task, first provide the symbol for effect size (*d*), and the formula for effect size that was provided in Chapter 13 of the *Student*

Guide. Following that, copy the formula again, this time substituting the actual values from the SAS output for the symbols. If the values that you are inserting in the formula have decimal places, use the same number of decimal places that these values actually had in the SAS output. Next, complete the steps that are necessary to solve for *d* by performing only one mathematical operation at each step. Your last step should provide your final value of *d*, rounded to two decimal places only.

4. Assume that you began with the following research question: "The purpose of this study was to ask a group of subjects to imagine how they would feel if they learned that their partners had had intercourse with someone else, and then determine whether male subjects would score higher than female subjects on a measure of psychological distress." The output created by your SAS program should be relevant to this research question.

   On a separate sheet of paper, prepare a report summarizing the analysis, using the format shown in the section titled "Summarizing the Results of the Analysis" from Chapter 13 of the *Student Guide*. You will have to modify the report so that it is relevant to the analysis that you are actually conducting. Underline each heading in this report. Remember to include information for sections A–M, along with a figure (a bar graph) that represents the results (section N).

What You Will Hand In

Hand in the following materials stapled together in this order:

1. A printout of your SAS program (including data), your SAS log, and your SAS output files (in that order).
2. A page titled "Computing the Index of Effect Size." This page should contain everything requested by step 3 in the section titled "Your Assignment" of this exercise.
3. A separate page containing your report summarizing the results of the analysis, including information for sections A–M and a figure (for section N).

Hint

If your SAS program ran correctly, your output should resemble the following output:

```
                              JANE DOE                                 1

                         The TTEST Procedure
                             Statistics

                         Lower CL            Upper CL  Lower CL
Variable  Class        N     Mean      Mean      Mean   Std Dev  Std Dev

DISTRESS  F           18   20.179  21.111    22.044    1.4071  1.8752
DISTRESS  M           18   22.914  23.944    24.975    1.5544  2.0714
DISTRESS  Diff (1-2)       -4.172  -2.833    -1.495    1.5981  1.9758

                              Statistics

                             Upper CL
    Variable  Class          Std Dev    Std Err    Minimum    Maximum

    DISTRESS  F               2.8112      0.442         18         25
    DISTRESS  M               3.1054     0.4882         20         28
    DISTRESS  Diff (1-2)      2.5886     0.6586


                               T-Tests

  Variable   Method          Variances       DF    t Value    Pr > |t|

  DISTRESS   Pooled            Equal         34      -4.30      0.0001
  DISTRESS   Satterthwaite     Unequal     33.7      -4.30      0.0001

                        Equality of Variances

  Variable    Method       Num DF     Den DF     F Value     Pr > F

  DISTRESS    Folded F         17         17        1.22     0.6862
```

References

Buunk, B.P., Angleitner, A., Oubaid, V., & Buss, D.M. (1996). Sex differences in jealousy in evolutionary and cultural perspective: Tests from the Netherlands, Germany, and the United States. *Psychological Science, 7*, 359-363.

Exercise 13.2: Effect of Interviewer Suspicion on Interviewee Nervousness

Overview

In this exercise, you will read about a fictitious study that was designed to determine whether individuals in a conversation will feel more nervous when the person interviewing them has been made to be suspicious. In this study, subjects are placed in dyads (pairs), and in half of the dyads, the interviewer is made to be suspicious of the interviewee. In the other half of the dyads, the interviewer is not made to be suspicious. You will be provided with fictitious data from this study, and you will prepare a SAS program that inputs this data. You will perform an independent-samples *t* test to determine whether there is a significant difference between the interviewees in the suspicious condition and the interviewees in the nonsuspicious condition with respect to their mean scores on a measure of nervousness. Finally, you will prepare an analysis report that summarizes your findings.

The Study

Suppose that you are a researcher in the field of communications. In your current investigation, you are studying suspicion. You want to determine whether suspicion can affect the dynamics between two people when they are involved in a conversation. You suspect that, when one speaker is suspicious of the other speaker, it can cause a number of responses on the part of the second speaker. One of these responses might be increased nervousness.

To test this, you conduct a simple experiment in which subjects are placed in dyads (pairs) and are asked to engage in a structured conversation. You manipulate the instructions so that, in half of the dyads, one speaker (the interviewer) is made to be suspicious of the other speaker (the interviewee). In the other half of the dyads, nothing is done to make anyone suspicious. At the end of the session, interviewees rate the extent to which they felt nervous. You then perform an independent-samples *t* test to determine whether there was a significant difference between the two groups with respect to the mean levels of self-reported nervousness.

Note: Although the study described here is fictitious, it was inspired by the actual investigation reported by Burgoon, et al. (1995).

Research question. The purpose of this study was to determine whether interviewees being interviewed by highly suspicious interviewers would display a higher level of nervousness, compared to interviewees being interviewed by less suspicious interviewers.

Research hypothesis. Interviewees being interviewed by highly suspicious interviewers will display a higher level of nervousness, compared to interviewees being interviewed by less suspicious interviewers.

Research Method. Suppose that you begin the study with 40 subjects, and you randomly group the subjects into pairs (also called dyads), leaving you with 20 pairs of subjects. Each pair is now randomly assigned to treatment conditions, so that 10 pairs of subjects are assigned to the "high-suspicion" condition and the other 10 pairs of subjects are assigned to the "low-suspicion" condition.

The study is conducted in three stages:

- **Stage 1.** Subjects are told that they are about to participate in a study dealing with personal values and beliefs. They complete a questionnaire that asks about a wide variety of their values and beliefs.
- **Stage 2.** The researchers explain that subjects will now pair up for purposes of the interview. In this session, one subject (the interviewer) will ask the second subject (the interviewee) a number of questions about the interviewee's values and beliefs.
- **Stage 3.** The interview is over and subjects complete an additional questionnaire. Some items on the questionnaire assess how nervous the subjects felt during the interview. Responses on this "nervousness" scale (completed by the interviewees) will serve as the criterion variable in the analysis.

Manipulating the predictor (independent) variable. The predictor variable in this study is "level of suspiciousness displayed by interviewers." It is necessary that you manipulate things so that the interviewers in the high-suspicion condition are more suspicious about their interviewees, compared to the interviewers in the low-suspicion condition.

To achieve this, you manipulate the instructions that are provided to interviewers after Stage 1:

- **High-suspicion condition.** After subjects have completed questionnaires in Stage 1, you meet individually with each interviewer and tell the interviewer that you have reason to believe that the interviewee gave untruthful answers to many items on the questionnaire. You urge the interviewer to try to obtain the truth during the interview conducted in Stage 2.
- **Low-suspicion condition.** After subjects have completed questionnaires in Stage 1, you meet individually with each interviewer and tell the interviewer that you believe that the interviewees gave honest answers to most of the items on the questionnaire. You tell the interviewer that he or she will probably also get truthful answers during the interview in Stage 2.

Measuring the criterion (dependent) variable. At Stage 3 (after the interview), all interviewees complete a number of scales. One of these is a 4-item scale designed to measure self-reported nervousness. Scores on the scale can range from 4 to 28, with higher values representing greater nervousness on the part of the interviewee.

Predictor and criterion variables in the analysis. The predictor variable in your study is "level of suspiciousness displayed by the interviewers." This is a nominal-level variable that consists of only two values: high suspicion and low suspicion. In your analysis, you will give this variable the SAS variable name SUS_GRP, which stands for "suspicion group." With this variable, you will use the symbol "L" to represent subjects in the low-suspicion group, and "H" to represent subjects in the high-suspicion group.

The criterion variable in your study is "level of nervousness reported by the interviewees." This variable will consist of the interviewees' scores on the 4-item measure of nervousness previously described. You can assume that this variable is on an interval scale, as it has approximately equal intervals, but no true zero point. In the analysis, you will give it the SAS variable name NERV.

Data Set To Be Analyzed

Table 13.E2.1 provides scores for the nervousness variable from the 20 interviewees who participated in the study.

Table 13.E2.1

Subject Scores for the Measure of Nervousness

Subject	Suspicion group[a]	Nervousness
01	L	17
02	L	20
03	L	19
04	L	19
05	L	15
06	L	16
07	L	18
08	L	17
09	L	16
10	L	18
11	H	19
12	H	21
13	H	20
14	H	18
15	H	20
16	H	19
17	H	20
18	H	17
19	H	22
20	H	23

[a] With the variable "Suspicion group," the value "L" identifies subjects in the low-suspicion condition, and the value "H" identifies subjects in the high-suspicion condition.

In Table 13.E2.1, you can see that Subject 01 has the value "L" in the "Suspicion group" column, indicating that this subject is in the low-suspicion condition. Subject 01 has the value "17" in the "Nervousness" column, indicating a score of 17 for the measure of nervousness. If you look at Subject 11, you can see that this subject has the value "H" in the "Suspicion group" column, indicating that the subject is in the high-suspicion condition. Subject 11 has the value "19" in the "Nervousness" column, indicating a score of 19 for the measure of nervousness.

Your Assignment

1. Create a SAS program that will input the data from Table 13.E2.1 and analyze it using PROC TTEST to determine whether there is a significant difference between the low-suspicion group and the high-suspicion group with respect to their scores for the nervousness variable. When you write this program, do the following:
   - Use the SAS programs presented in Chapter 13 of the *Student Guide* as models.
   - Use the SAS variable name SUB_NUM to represent subject numbers.
   - Use the SAS variable name SUS_GRP to represent your "level of suspicion" predictor variable. When you type this SAS variable name, use an underscore (_) and not a hyphen (-).
   - When you type values for the SUS_GRP variable, use the value "L" to represent subjects in the low-suspicion condition, and the value "H" to represent subjects in the high-suspicion condition (as in Table 13.E2.1).
   - Use the SAS variable name NERV to represent subject scores for the nervousness variable.
   - Type your full name in the TITLE1 statement, so that it will appear in the output.
2. Submit the program for analysis, and, if necessary, correct it so that it runs without errors.
3. At the top of a clean sheet of paper, write the title "Computing the Index of Effect Size." On this sheet of paper, compute the index of effect size (d)

for the current analysis. You will have to do this by hand; see Chapter 13 of the *Student Guide* for details about computing the index of effect size.

In performing this task, first provide the symbol for effect size (*d*), and the formula for effect size that was provided in Chapter 13 of the *Student Guide*. Following that, copy the formula again, this time substituting the actual values from the SAS output for the symbols. If the values that you are inserting in the formula have decimal places, use the same number of decimal places that these values actually had in the SAS output. Next, complete the steps that are necessary to solve for *d* by performing only one mathematical operation at each step. Your last step should provide your final value of *d*, rounded to two decimal places only.

4. Assume that you began with the following research question: "The purpose of this study was to determine whether interviewees being interviewed by highly suspicious interviewers would display a higher level of nervousness, compared to interviewees being interviewed by less suspicious interviewers." The output created by your SAS program should be relevant to this research question.

   On a separate sheet of paper, prepare a report summarizing the analysis, using the format shown in the section titled "Summarizing the Results of the Analysis" from Chapter 13 of the *Student Guide*. You will have to modify the report so that it is relevant to the analysis that you are actually conducting. Underline each heading in this report. Remember to include information for sections A–M, along with a figure (a bar graph) that represents the results (section N).

What You Will Hand In

Hand in the following materials stapled together in this order:

1. A printout of your SAS program (including data), your SAS log, and your SAS output files (in that order).
2. A page titled "Computing the Index of Effect Size." This page should contain everything requested by step 3 in the section titled "Your Assignment" of this exercise.

3. A separate page containing your report summarizing the results of the analysis, including the information for sections A–M and a figure (for section N).

Hint

If your SAS program ran correctly, your output should resemble the following output:

```
                              JOHN  DOE                                      1
                        The TTEST Procedure
                             Statistics

                            Lower CL           Upper CL Lower CL
Variable  Class       N        Mean     Mean       Mean  Std Dev  Std Dev

NERV      H          10      18.618     19.9     21.182   1.2326    1.792
NERV      L          10      16.369     17.5     18.631   1.0876   1.5811
NERV      Diff (1-2)         0.8123      2.4     3.9877   1.2769   1.6898

                             Statistics

                           Upper CL
  Variable  Class           Std Dev     Std Err     Minimum     Maximum

  NERV      H                3.2714      0.5667          17          23
  NERV      L                2.8865         0.5          15          20
  NERV      Diff (1-2)        2.499      0.7557

                              T-Tests

Variable    Method            Variances       DF     t Value    Pr > |t|

NERV        Pooled            Equal           18        3.18      0.0052
NERV        Satterthwaite     Unequal       17.7        3.18      0.0053

                       Equality of Variances

   Variable     Method        Num DF      Den DF     F Value      Pr > F

   NERV         Folded F           9           9        1.28      0.7153
```

Reference

Burgoon, J.K., Buller, D.B., Dillman, L., & Walther, J.B. (1995). Interpersonal deception IV. Effects of Suspicion on perceived communication and nonverbal behavior dynamics. *Human Communication Research, 22*, 163-196.

EXERCISES 14

Exercises for Chapter 14: Paired-Samples *t* Test

Exercise 14.1: Perceived Problem Seriousness as a Function of Time of Day

Overview

Do people view their personal problems as being more serious at one time of day (for example, in the afternoon) compared to another time of day (for example, in the morning)? The present study is designed to answer this question. In this fictitious study, you will have a group of people complete a brief scale that assesses perceived problem seriousness. They will complete the scale at two points in time during the day: once in the morning, and once in the afternoon. You will then use a paired-samples *t* test to analyze their responses to determine whether perceived problem seriousness scores obtained in the afternoon are significantly higher than those obtained in the morning. Finally, you will prepare an analysis report that summarizes your findings.

Note: Although the investigation reported here is fictitious, it was inspired by the actual study reported by Thayer (1987).

The Study

Research question. The purpose of this study was to determine whether there is a difference between mean perceived problem seriousness scores obtained during the afternoon and those obtained in the morning.

Research hypothesis. Mean perceived problem seriousness scores obtained during the afternoon will be higher (rated as more serious) than those obtained in the morning.

Research Method

Measuring the criterion variable. The criterion variable in your study is perceived problem seriousness. Assume that you ask a group of 16 subjects to each identify some personal problem that they are dealing with. The problem might involve difficulties with parents, poor academic performance, problems in romantic relationships, or any other problem. A given subject is asked to provide ratings of how serious his or her problem is by responding to the following items:

```
1.   How serious do you believe your problem is?

  Not at all serious  1  2  3  4  5  6  7  Extremely serious


2.   How difficult do you believe it will be to solve your
     problem?

 Not difficult at all 1  2  3  4  5  6  7  Extremely difficult


3.   How likely is it that you will solve your problem?

           Very likely  1  2  3  4  5  6  7  Not likely at all
```

Subjects respond to each item by circling a number between 1 and 7. When finished, you compute a given subject's score on this "perceived problem seriousness" scale by summing the numbers circled by that subject. Scores on the scale may range from a low of 3 (if the subject circled "1" for each item) to a high of 21 (if the subject circled "7" for each item). The higher the score, the more serious the problem is perceived to be by the subject.

Procedure. You ask subjects to complete this scale twice each day: once late in the morning (around 11:00 AM), and again late in the afternoon (around 4:00 PM). Subjects do this each day for two weeks, and at the end of the two-week period, you compute only two scores for each subject: the subject's average morning score, and the subject's average afternoon score (to avoid decimal places, you round each average score to the nearest whole number).

With the data gathered, you can now compare the average perceived problem seriousness scores obtained in the afternoon to the average perceived problem seriousness scores obtained in the morning. This will help you understand whether these perceptions vary systematically according to the time of day.

Predictor and criterion variables in the analysis. The predictor variable in your study is "time of day." This is a nominal-level variable that consists of only two conditions: morning and afternoon.

The criterion variable in your study is "perceived problem seriousness," which is the subjects' scores on the 3-item measure previously described. You may assume that this variable is on an interval scale, as it has approximately equal intervals, but no true zero point. In your analysis, you will use the SAS variable name MORNING to represent scores obtained in the morning, and the SAS variable name AFTNOON to represent scores obtained in the afternoon.

Data Set to be Analyzed

Table 14.E1.1 presents the scores for the criterion variable for each of the 16 subjects in your study.

Table 14.E1.1

Data from the Problem Seriousness Study

Subject number	Perceived problem seriousness scores: After-noon	Morning
01	16	13
02	14	14
03	12	9
04	16	12
05	13	12
06	11	13
07	17	13
08	10	10
09	14	11
10	15	11
11	14	8
12	13	11
13	15	11
14	13	8
15	13	10
16	11	10

Each horizontal row in the preceding table presents the perceived problem seriousness scores obtained for a specific subject under the two conditions. For example, the first row presents results for Subject 01, who provided a mean problem seriousness score of 16 under the afternoon condition, and a problem seriousness score of 13 under the morning condition. Subject 02 provided a score of 14 under the afternoon condition, and a score of 14 under the morning condition. Data for the remaining subjects may be interpreted in the same way.

Your Assignment

1. Create a SAS program that will input the data set in Table 14.E1.1 and analyze it using PROC MEANS and PROC TTEST to perform a paired-samples *t* test. Specifically, you will determine whether the mean score

obtained under the afternoon condition is significantly different from the mean score obtained under the morning condition.

When you write this program, do the following:

- Use the SAS programs presented in Chapter 14 of the *Student Guide* as models.
- When you type your data, type all of the variables from Table 14.E1.1. You *should* type the subject number variable that appears as the first column in the table. Give this variable the SAS variable name SUB_NUM. When typing this SAS variable name, make sure you use an underscore (_) and not a hyphen (-). When typing subject numbers such as "01," make sure you type a zero (0) and not the letter O.
- Use the SAS variable name AFTNOON to represent perceived problem difficulty scores obtained during the afternoons (do *not* use the SAS variable name AFTERNOON). Use the SAS variable name MORNING to represent problem seriousness scores obtained during the mornings.
- As was done in the example programs in the *Student Guide*, use PROC MEANS to obtain descriptive statistics (means and standard deviations and so forth) for the two criterion variables, and use PROC TTEST to actually perform the paired-samples *t* test. Include both PROCs within the same program.
- Type your full name in the TITLE1 statement, so that it will appear in each page of output.

2. Submit the program for analysis, and, if necessary, correct it so that it runs without errors.
3. At the top of a clean sheet of paper, write the title "Computing the Index of Effect Size." On this sheet of paper, compute the index of effect size (*d*) for the current analysis. You will have to do this by hand; see Chapter 14 of the *Student Guide* for details about computing the index of effect size.

   In performing this task, first provide the symbol for effect size (*d*), and the formula for effect size that was provided in Chapter 14 of the *Student Guide*. Following that, copy the formula again, this time substituting the actual values from the SAS output for the symbols. If the values that you are inserting in the formula have decimal places, use the same number of decimal places that these values actually had in the SAS output. Next, complete the steps that are necessary to solve for *d* by performing only one mathematical operation at each step. Your last step should provide your final value of *d*, rounded to two decimal places only.

4. Assume that you began with the following research question: "The purpose of this study was to determine whether there is a difference between mean perceived problem seriousness scores obtained during the afternoon versus those obtained in the morning." The output created by your SAS program should be relevant to this research question.

   On a separate sheet of paper, prepare a report summarizing the analysis, using the format shown in the section titled "Summarizing the Results of the Analysis" from Chapter 14 of the *Student Guide*. You will have to modify the report so that it is relevant to the present analysis.

   The research hypothesis stated toward the beginning of this exercise is a directional hypothesis. It states "Mean perceived problem seriousness scores obtained during the afternoon will be higher (rated as more serious) than those obtained in the morning." You should include this research hypothesis in your report. However, your analysis report should present your statistical null hypothesis and statistical alternative hypotheses as *nondirectional* (two-tailed) hypotheses . You will do this so that your region of rejection is not concentrated in only one tail of the sampling distribution.

   Underline each heading in this report. Remember to include information for sections A–M (as shown in Chapter 14 of the *Student Guide*) along with a figure (bar graph) for section N. If numbers in your report have decimal places, round to two decimal places (except for the *p* value, which should be rounded to four decimal places).

What You Will Hand In

Hand in the following materials stapled together in this order:

1. A printout of your SAS program (including data), your SAS log, and your SAS output files.

2. A page titled "Computing the Index of Effect Size." This page should contain everything requested by step 3 in the section titled "Your Assignment" of this exercise.
3. A separate page containing your report summarizing the results of the analysis, including information for sections A–M and a figure (for section N).

Hint

If your SAS program ran correctly, your output should resemble the following output. Your results from PROC MEANS and your results from PROC TTEST will probably appear on two separate pages of output.

```
                          JOHN DOE                                  1

                     The MEANS Procedure

Variable     N          Mean        Std Dev       Minimum         Maximum
--------------------------------------------------------------------------
AFTNOON     16    13.5625000      1.9653244    10.0000000      17.0000000
MORNING     16    11.0000000      1.7888544     8.0000000      14.0000000
--------------------------------------------------------------------------
```

```
                          JOHN DOE                             2

                    The TTEST Procedure

                         Statistics

                      Lower CL              Upper CL  Lower CL
Difference         N      Mean      Mean       Mean    Std Dev Std Dev

AFTNOON - MORNING  16   1.4453   2.5625     3.6797     1.5488  2.0966

                          Statistics

                        Upper CL
    Difference           Std Dev     Std Err     Minimum     Maximum

    AFTNOON - MORNING     3.2449      0.5242          -2           6


                           T-Tests

           Difference               DF     t Value     Pr > |t|

           AFTNOON - MORNING        15        4.89       0.0002
```

Reference

Thayer, R.E. (1987). Problem perception, optimism, and related states as a function of time of day (diurnal rhythm) and moderate exercise: Two arousal systems in interaction. *Motivation and Emotion, 11,* 19-34.

Exercise 14.2: Effect of Functional Family Therapy on Probation Compliance for Juvenile Delinquents

Overview

This exercise describes a fictitious study designed to determine whether functional family therapy can have a positive effect on the behavior of juvenile delinquents. In this study, 18 juvenile delinquents are assigned to a "therapy" condition. Each subject in this condition receives probation along with functional family therapy. Each subject in the therapy condition is paired with another juvenile delinquent in the "control" condition. Pairs are formed by using a matching procedure. The 18 subjects in the control condition receive probation, but no family therapy. At the end of a 24-month period, subjects in both conditions are rated on probation compliance. Because each subject in the therapy condition is paired with a similar subject in the control condition, you will analyze the data using a paired-samples *t* test to determine whether the subjects in the therapy condition displayed higher levels of compliance, compared to those in the control condition. You will also prepare an analysis report in which you summarize your analysis and results.

Note: Although the investigation reported here is fictitious, it was inspired by the actual study reported by Gordon et al. (1995).

The Study

Research question. The purpose of this study was to determine whether there is a difference between delinquents who receive probation along with functional family therapy and delinquents who receive probation only, with respect to their mean scores on a measure of probation compliance.

Research hypothesis. Delinquents who receive probation along with functional family therapy will demonstrate higher mean scores on probation compliance, compared to delinquents who receive probation only.

Research Method

Suppose that you are a researcher in the area of criminal justice. You are attempting to identify interventions that may be helpful in rehabilitating juveniles who commit crimes. In your current project, you are assessing the effectiveness of *functional family therapy*, an intervention that uses ideas from social learning theory and systems theory.

To assess the effectiveness of this intervention, you conduct a study using two groups of subjects:

- **Therapy group**. The therapy group consists of 18 female adolescents convicted of crimes in juvenile court. At the time that they are placed on probation, they and their parents begin five months of treatment in functional family therapy.
- **Control group**. The control group consists of 18 female adolescents convicted of crimes in juvenile court. These subjects are placed on the same type of probation as the therapy group, but they do not receive functional family therapy.

The matching procedure. A matching procedure is used to place subjects into pairs. Each of the 18 subjects in the therapy group is paired with a similar subject in the control group. Matching is done on the basis of age, race, socioeconomic status, and type of crime. For example:

- A 14-year-old Caucasian from a low-income group who committed vandalism and is assigned to the therapy group will be paired with a subject from the control group who is a 15-year-old Caucasian from a low-income group who committed vandalism.
- A 16-year-old Asian-American from a middle-income group who committed shoplifting and is assigned to the therapy group will be paired with a subject from the control group who is a 16-year-old Asian-American from a middle-income group who committed shoplifting.

Although there are actually 36 subjects in your study, they are grouped into only 18 *pairs* of subjects, and each pair will serve as an observation. This means that, for your study, $N = 18$. Because of this matching procedure, you will be able to analyze your data using a paired-samples *t* test, rather than an independent-samples *t* test.

At the end of the 24-month period following their sentencing, subjects in both groups are rated on probation compliance. Here, "probation compliance" refers to the extent to which the delinquent has adhered to the rules and guidelines set down by the judge at the time of her sentencing.

The probation compliance ratings are made by the delinquent's parents and probation officers. They are made using a multiple-item questionnaire containing statements such as "She regularly attends school." Each subject receives a score on probation compliance that may range from zero through 20, with higher scores indicating higher levels of compliance.

Predictor and criterion variables in the analysis. The predictor variable in your study is "type of intervention." This is a nominal-level variable that consists of only two conditions: (a) probation along with functional family therapy, and (b) probation alone.

The criterion variable in your study is "probation compliance," which is calculated from the subject's ratings on the measure previously described. You may assume that this variable is on an interval scale, as it has approximately equal intervals, but no true zero point. In your analysis, you will use the SAS variable name THERAPY to represent compliance scores for the delinquents in the therapy condition, and the SAS variable name CONTROL to represent compliance scores for the delinquents in the control condition.

Data Set to be Analyzed

Table 14.E2.1 presents the scores for the criterion variable for each of the 18 pairs of subjects in your study.

Table 14.E2.1

Data from the Probation Compliance Study

	Probation compliance scores	
Pair number	Therapy condition	Control condition
01	15	13
02	16	10
03	16	15
04	17	17
05	12	08
06	14	11
07	15	07
08	14	13
09	16	17
10	13	15
11	16	16
12	18	14
13	17	14
14	16	14
15	17	12
16	15	10
17	14	15
18	14	16

Table 14.E2.1 is somewhat different from the data table that appeared in Chapter 14 of the *Student Guide*. In Table 14.E2.1, there is no column headed "Subject number;" instead, this has been replaced with a column headed "Pair number." This is because, in the present analysis, the unit of analysis is the individual *pair* of subjects, not the individual subject.

Scores for the criterion variable appear below the heading "Probation compliance scores." Below the heading "Therapy condition" are the compliance scores for the 18 subjects who received probation along with functional family therapy. Below the heading "Control condition" are the compliance scores for the 18 subjects who received probation only.

A given horizontal row in Table 14.E2.1 provides data for two subjects who were paired together. For example:

- In the row for pair number "01," you can see that the subject in the therapy condition received a score of 15 for compliance, while her counterpart in the control condition (the girl with whom she was paired) received a score of 13.
- In the row for pair number "02," you can see that the subject in the therapy condition received a score of 16 for compliance, while her counterpart in the control condition (the girl with whom she was paired) received a score of 10.

Each of the remaining rows may be interpreted in the same way.

Your Assignment

1. Create a SAS program that will input the data set in Table 14.E2.1 and analyze it using PROC MEANS and PROC TTEST to perform a paired-samples *t* test. Specifically, you will determine whether the mean score obtained under the therapy condition is significantly different from the mean score obtained under the control condition.

   When you write this program, do the following:

   - Use the SAS programs presented in Chapter 14 of the *Student Guide* as models.
   - When you type your data, type all of the variables from Table 14.E2.1.You *should* type the pair number variable that appears as the first column in the table. Give this variable the SAS variable name PAIR_NUM. When typing this SAS variable name, make sure you use an underscore (_) and not a hyphen (-). When typing numbers such as "01," make sure you type a zero (0) and not the letter O.
   - Use the SAS variable name THERAPY to represent probation compliance scores obtained under the therapy condition. Use the SAS variable name CONTROL to represent compliance scores obtained under the control condition.
   - As was done in the example programs in the *Student Guide*, use PROC MEANS to obtain descriptive statistics (means and standard deviations and so forth) for the two criterion variables, and use PROC TTEST to

actually perform the paired-samples *t* test. Include both PROCs within the same program.

- Type your full name in the TITLE1 statement so that it will appear in each page of output.

2. Submit the program for analysis, and, if necessary, correct it so that it runs without errors.

3. At the top of a clean sheet of paper, write the title "Computing the Index of Effect Size." On this sheet of paper, compute the index of effect size (*d*) for the current analysis. You will have to do this by hand; see Chapter 14 of the *Student Guide* for details about computing the index of effect size.

   In performing this task, first provide the symbol for effect size (*d*), and the formula for effect size that was provided in Chapter 14 of the *Student Guide*. Following that, copy the formula again, this time substituting the actual values from the SAS output for the symbols. If the values that you are inserting in the formula have decimal places, use the same number of decimal places that these values actually had in the SAS output. Next, complete the steps that are necessary to solve for *d* by performing only one mathematical operation at each step. Your last step should provide your final value of *d*, rounded to two decimal places only.

4. Assume that you began with the following research question: "The purpose of this study was to determine whether there is a difference between delinquents who receive probation along with functional family therapy and delinquents who receive probation only, with respect to their mean scores on a measure of probation compliance." The output created by your SAS program should be relevant to this research question.

   On a separate sheet of paper, prepare a report summarizing the analysis, using the format shown in the section titled "Summarizing the Results of the Analysis" from Chapter 14 of the *Student Guide*. You will have to modify the report so that it is relevant to the present analysis.

   The research hypothesis stated toward the beginning of this exercise is a directional hypothesis. It states "Delinquents who receive probation along with functional family therapy will demonstrate higher mean scores on probation compliance, compared to delinquents who receive probation only." You should include this research hypothesis in your report.

However, your analysis report should present your statistical null hypothesis and statistical alternative hypotheses as *nondirectional* (two-tailed) hypotheses . You will do this so that your region of rejection is not concentrated in only one tail of the sampling distribution.

Underline each heading in this report. Remember to include information for sections A–M (as shown in Chapter 14 of the *Student Guide*) along with a figure (bar graph) for section N. If numbers in your report have decimal places, round to two decimal places (except for the *p* value, which should be rounded to four decimal places).

What You Will Hand In

Hand in the following materials stapled together in this order:

1. A printout of your SAS program (including data), your SAS log, and your SAS output files.
2. A page titled "Computing the Index of Effect Size." This page should contain everything requested by step 3 in the section titled "Your Assignment" of this exercise.
3. Your report summarizing the results of the analysis, including information for sections A–M and a figure (for section N).

Hint

If your SAS program ran correctly, your output should resemble the following output. Your results from PROC MEANS and your results from PROC TTEST will probably appear on two separate pages of output.

```
                              JANE DOE                                1

                         The MEANS Procedure

Variable       N          Mean       Std Dev       Minimum       Maximum
-------------------------------------------------------------------------
THERAPY       18    15.2777778     1.5645167    12.0000000    18.0000000
CONTROL       18    13.1666667     2.9555531     7.0000000    17.0000000
-------------------------------------------------------------------------
```

```
                         JANE DOE                             2

                    The TTEST Procedure

                        Statistics

                    Lower CL          Upper CL  Lower CL
Difference         N    Mean     Mean     Mean   Std Dev  Std Dev

THERAPY - CONTROL 18  0.6852  2.1111   3.5371    2.1517   2.8674

                        Statistics

                         Upper CL
    Difference           Std Dev    Std Err    Minimum    Maximum

    THERAPY - CONTROL     4.2987     0.6759         -2          8


                         T-Tests

         Difference             DF    t Value    Pr > |t|

         THERAPY - CONTROL      17       3.12      0.0062
```

Reference

Gordon, D.A., Graves, K., & Arbuthnot, J. (1995). The effect of functional family therapy for delinquents on adult criminal behavior. *Criminal Justice and Behavior, 22*, 60-73.

Exercises for Chapter 15: One-Way ANOVA with One Between-Subjects Factor

Exercise 15.1: The Effect of Misleading Suggestions on the Creation of False Memories

Overview

In this exercise, you will read about a study in which witnesses are subjected to a form of suggestive questioning in order to determine whether it can lead to the formation of false memories. Here, **false memories** are defined as erroneous memories for events that never took place.

The Study

This study is designed to determine whether the number of "misleading suggestions" made during questioning is related to the witness' confidence that some nonexistent event actually took place. Suppose that you are conducting this experiment, and you predict that the greater the number of misleading suggestions, the greater the witness' confidence that the nonexistent event actually happened.

In this study, subjects will watch a 5-minute videotape in which two young men burglarize a home. After viewing the videotape, subjects are exposed to a varying number of misleading suggestions. The misleading suggestions are designed to make subjects believe that one of the thieves wore gloves when, in reality, he did not. In the study, some subjects are exposed to zero misleading suggestions, some are exposed to two, and some are exposed to four. One week later, subjects rate their confidence that the thief wore gloves.

You will analyze the data to determine whether the number of misleading suggestions experienced by the subjects is positively related to their confidence that the thief wore gloves. If you find such a relationship, the results will be consistent with the hypothesis that false memories can be created, and that the greater the number of misleading suggestions, the greater the confidence placed on the false memory.

Note: Although the study reported here is fictitious, it was inspired by the actual study reported by Zaragoza and Mitchell (1996).

Research question. The purpose of this study was to determine whether there was a relationship between (a) the number of exposures to misleading suggestions made during questioning and (b) subject confidence that the thief in the video wore gloves.

Research hypothesis. There will be a positive relationship between the number of exposures to misleading suggestions and subject confidence that the thief wore gloves.

Research Method

Subjects. Your study includes a total of 21 subjects, each of whom are randomly assigned to one of three treatment conditions.

The eyewitness event. Regardless of experimental condition, all subjects in your study view the same 5-minute videotape of a burglary. In the video, two young people burglarize a home and are involved in a car chase with the police.

Manipulating the predictor (independent) variable: Number of exposures to misleading suggestions. Immediately after viewing the video, each subject completes a 36-item questionnaire. This questionnaire contains 36 YES/NO items such as:

> At the beginning of the video, a blonde-haired youth wearing a leather jacket entered the home. Did he gain entry through a window?

This questionnaire is used to manipulate the study's predictor (independent) variable: the number of exposures to misleading suggestions. In this investigation, a **misleading suggestion** is operationalized as a question that presupposes the existence of events or objects that were not actually in the video (although they were plausible). For example, assume that your video does show that the blonde-haired burglar wore a leather jacket, but did not wear gloves. Assume further that you wish to provide a given subject with a misleading suggestion that the burglar *did* wear gloves. To do this, you include the following item in your post-video questionnaire:

> At the beginning of the video, a blonde-haired youth wearing a leather jacket and gloves entered the home. Did he gain entry through a window?

Notice that the preceding question has been modified so that it now presupposes that the burglar was wearing gloves (even though, in reality, he was not). In this way, you have now provided your subject with a misleading suggestion.

You manipulate this independent variable by varying the number of misleading suggestions that appear in the post-video questionnaire. Subjects assigned to the different treatment conditions see different numbers of misleading suggestions:

- The seven subjects assigned to the zero-exposures condition are exposed to no misleading suggestions.
- The seven subjects assigned to the two-exposures condition are exposed to two misleading suggestions.
- The seven subjects assigned to the four-exposures condition are exposed to four misleading suggestions.

Research design. The research design for the present study is presented in Figure 15.E1.1:

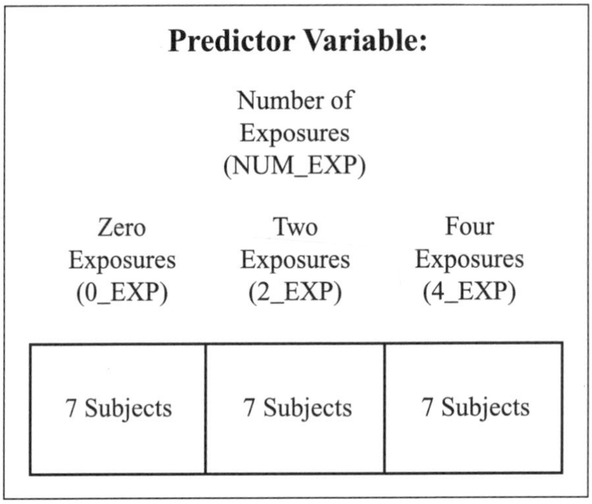

Figure 15.E1.1: Research Design Used in the False Memories Study.

Measuring the criterion (dependent) variable. The criterion (dependent) variable in your study is the subjects' confidence that the blonde-haired thief wore gloves in the video. As you know, this thief did not wear gloves in the video, so it is reasonable to assume that those subjects who are confident that he did wear gloves are displaying a false memory.

To assess this dependent variable, you ask each subject to return for another session one week after they watched the video. In this second session, you ask them to complete an 80-item questionnaire in which they indicate whether they believe that certain events occurred in the burglary video that they had previously viewed.

The questionnaire lists a number of possible facts and events, such as:

- The thief stole a lamp.
- The house lights were all off.
- The police car's siren was on.
- The blonde-haired thief wore gloves.

For each item, subjects indicate how confident they are that the video actually portrayed this event. They make their ratings using a 7-point response format in which "7" represents "Definitely Yes" and "1" represents "Definitely No."

For four of the items in the questionnaire, subjects indicate their confidence that "The blonde-haired thief wore gloves" (this idea will be worded somewhat differently in each of the four different items). When you analyze your data, you focus on subject responses to these four items only, and ignore the other 76 items in the questionnaire.

For each subject, you compute his or her average response to the four items dealing with the idea that the thief wore gloves. The resulting mean serves as that subject's score for the dependent variable. For a given subject, this score could range from a low of 1.00 (indicating that subject definitely does not believe that the thief wore gloves), to a high of 7.00 (indicating that subject definitely does believe that the thief wore gloves). Because you are averaging responses to four items, individual scores for this dependent variable could also be fractional values such as 1.75, 3.50, 4.25, and so forth.

The Predictor and Criterion Variables in the Analysis. The predictor variable in your study is "number of exposures to misleading suggestions." This is a ratio-level variable that consists of three conditions: (a) zero exposures, (b) two exposures, and (c) four exposures. In your analysis, you will use the SAS variable name "NUM_EXP" to represent group membership for this predictor.

The criterion variable in your study is "confidence that the thief wore gloves:" subject ratings for the measure described in the previous section of this exercise. You may assume that this variable is on an interval scale, as it has approximately equal intervals, but no true zero point. In your analysis, you will use the SAS variable name "CONFID" to represent this variable.

Data Set to be Analyzed

Table 15.E1.1 presents the data set that you will analyze in this exercise.

Table 15.E1.1

Data from the False Memories Study

Subject	Number of exposures	Confidence score
01	0_EXP	1.00
02	0_EXP	1.25
03	0_EXP	2.00
04	0_EXP	1.75
05	0_EXP	2.75
06	0_EXP	3.25
07	0_EXP	4.50
08	2_EXP	2.75
09	2_EXP	3.00
10	2_EXP	4.00
11	2_EXP	5.50
12	2_EXP	5.50
13	2_EXP	6.25
14	2_EXP	6.75
15	4_EXP	4.75
16	4_EXP	3.00
17	4_EXP	4.75
18	4_EXP	5.75
19	4_EXP	7.00
20	4_EXP	6.00
21	4_EXP	3.00

Each row in Table 15.E1.1 (running horizontally) provides data for a single subject. For example, the first subject has the value "01" for Subject, the value "0_EXP" for Number of exposures, and a value of "1.00" for Confidence score.

The first column (running vertically) in Table 15.E1.1 is headed "Subject." This column assigns a unique subject number to each participant in the study.

The second column (running vertically) in Table 15.E1.1 is headed "Number of exposures." This column indicates the group to which each subject was assigned for the predictor variable. The first seven subjects are identified with the value "0_EXP," which represents "zero exposures." The next seven subjects are identified with the value "2_EXP," which represents "two exposures." The last seven subjects are identified with the value "4_EXP," which represents "four exposures."

The third column in Table 15.E1.1 is headed "Confidence score." This column provides each subject's score for the dependent variable: their confidence that the thief wore gloves. You will remember that this measure could range from a low of 1.00 (meaning that the subject definitely does not believe that the thief wore gloves), to a high of 7.00 (indicating that subject definitely does believe that the thief wore gloves).

When you write your SAS program, you will type your data as it appears in Table 15.E1.1; you will not need to reorganize the data.

Your Assignment

1. Create a SAS program that will input the data set in Table 15.E1.1. Use PROC GLM to perform a one-way ANOVA with one between-subjects factor. In this analysis, the criterion variable should be "confidence that the thief wore gloves," and the predictor variable should be "number of exposures to misleading suggestions."

   When you write this program, do the following:

   - Use the SAS programs presented in Chapter 15 of the *Student Guide* as models.
   - When typing your data, type all of the variables from Table 15.E1.1. You *should* type the subject number variable that appears as the first column in this table. Give this variable the SAS variable name SUB_NUM. When typing this SAS variable name, make sure you use an underscore (_) and not a hyphen (-). When typing subject numbers such as "01," make sure you type a zero (0) and not the letter O.
   - Use the SAS variable name "NUM_EXP" to represent the predictor variable (number of exposures).
   - Use the SAS variable name CONFID to represent subject scores for the criterion variable (confidence that the thief wore gloves).
   - When typing your data, use the values that are presented in Table 15.E1.1. For example, use the values "0_EXP," "2_EXP," and

"4_EXP" to define conditions for the predictor variable. When typing the value "0_EXP," make sure you type a zero (0) and not the letter O.

- Include two MEANS statements in your program: one to request means and standard deviations, and one to request the Tukey multiple comparison procedure (request the same options that were requested in Chapter 15 of the *Student Guide*).
- Type your full name in the TITLE1 statement, so that it will appear in the output.

2. Submit the program for analysis, and, if necessary, correct it so that it runs without errors.
3. On a clean sheet of paper, prepare an ANOVA summary table that summarizes the results of your analysis. In doing this, follow the directions provided in the *Student Guide*, and use the ANOVA summary tables presented in Chapter 15 of the *Student Guide* as models. Be sure to (a) give your table a table number, (b) give your table a title appropriate for this study, (c) include a *p* value note at the bottom of the table that is appropriate for the level of significance observed in your output, and (d) include the R^2 value. If values in the table contain decimals, report those values to two decimal places.
4. On a separate sheet of paper, prepare a table that provides the differences between means, the confidence intervals for differences between means, and the results of the Tukey tests. This should be a single table, similar to the table presented in the section titled "Prepare a table that presents the results of the Tukey tests and the confidence intervals" from Chapter 15 of the *Student Guide*. Be sure to (a) give your table a table number, (b) give your table a title appropriate for this study, and (c) include a *p* value note at the bottom of the table that is appropriate for the level of significance observed in your output. If values in the table contain decimals, report those values to three decimal places.
5. Assume that you begin with the following research question and research hypothesis:
   - **Research question**. The purpose of this study was to determine whether there was a relationship between (a) the number of exposures

to misleading suggestions made during questioning and (b) subject confidence that the thief in the video wore gloves.

- **Research hypothesis**. There will be a positive relationship between the number of exposures to misleading suggestions and subject confidence that the thief wore gloves. Specifically, it is predicted that (a) subjects exposed to four suggestions will demonstrate a higher level of confidence than subjects exposed to two or zero suggestions, and (b) subjects exposed to two suggestions will demonstrate a higher level of confidence than subjects exposed to zero suggestions.

On a separate sheet of paper, prepare an analysis report to summarize your findings, using the format shown in the section titled "Analysis Report for the Aggression Study" from Chapter 15 of the *Student Guide*. You will have to modify that report so that it is relevant to your analysis.

Underline each heading in your report. Remember to include information for sections A–N (as shown in Chapter 15 of the *Student Guide*) with a figure (for section O). If values in your report contain decimals, round to two decimal places (except for the p value, which should be reported to four decimal places).

6. For section O of step 5, prepare a bar chart that plots the means for the three conditions in your study. In doing this, use the figures presented in Chapter 15 of the *Student Guide* as models.

What You Will Hand In

Hand in the following materials stapled together in this order:

1. A printout of your SAS program (including data), your SAS log, and your SAS output files (your pages of output should be in the correct order).
2. A copy of the ANOVA summary table that you have prepared as a part of your report.
3. A copy of the table that provides the differences between means, the confidence intervals for differences between means, and the results of the Tukey tests. This should be a single table that you have prepared as a part of your report.

4. A copy of your report summarizing the results of the analysis.
5. A figure (bar chart) in which you plot the means for the three conditions of your study. This figure should be included with your report summarizing the results of the analysis.

Hint

If your SAS program ran correctly, your output should resemble the following excerpts from the output of a correctly written program. Remember that your program will produce more output than shown in these excerpts.

```
Source         DF     Type III SS     Mean Square    F Value    Pr > F

NUM_EXP         2     29.18452381     14.59226190       6.97    0.0057
```

```
                        The GLM Procedure

        Level of                 ------------CONFID-----------
        NUM_EXP         N              Mean            Std Dev

        0_EXP           7        2.35714286         1.23201345
        2_EXP           7        4.82142857         1.57925540
        4_EXP           7        4.89285714         1.50594062
```

Reference

Zaragoza, M. S., & Mitchell, K. J. (1996). Repeated exposure to suggestion and the creation of false memories. *Psychological Science*, *7*, 294-300.

Exercise 15.2: The Effect of News Source Credibility on Voter Reactions to Political Scandal

The Study

Overview. In this study, subjects read a description about a congressman who is said to be involved in a scandal. After reading the description, subjects are asked to play the role of constituents in the congressman's district and to rate the extent to which they approve of the congressman's overall performance.

Subjects are randomly assigned to one of three treatment conditions, and the description of the congressman's situation varies according to condition. Subjects in one condition are told that the news source implicating the congressman in the scandal is a low-credibility news source (one that is not necessarily reliable). Subjects in the other two conditions are told that the news source is either a moderate-credibility source or a high-credibility source. As the researcher, you want to determine whether the credibility of the news source affects the subjects' approval ratings for the congressman.

Research question. The purpose of this study is to determine whether there is a significant relationship between (a) the credibility of a news source and (b) voter (subject) approval ratings for a congressman implicated in a scandal.

Research hypothesis. There will be a negative relationship between news source credibility and voter approval ratings for a congressman implicated in a scandal. Specifically, it is predicted that (a) subjects exposed to the high-credibility condition will display lower approval ratings than subjects exposed to the moderate- or low-credibility conditions, and (b) subjects exposed to the moderate-credibility condition will display lower approval ratings than subjects exposed to the low-credibility condition.

Research Method

Subjects. Your study includes a total of 27 subjects, each of whom are randomly assigned to one of three treatment conditions.

Profiles of office-holders. Regardless of the experimental condition, all subjects in your study are asked to read profiles of six politicians holding a variety of offices. Each profile describes the office held, provides background information for the politician (for example, education and work history), describes the politician's achievements while in office, and provides other miscellaneous information.

After reading the profile for a given politician, subjects complete a 4-item "voter approval" scale for that office-holder. Subjects play the role of voters in the district represented by that office holder, and rate the extent to which they approve of the politician's overall performance. The scale they complete includes items such as "I approve of this office holder's performance." They respond to each item using a 7-point response format in which "1" represents "Disagree Very Strongly" and "7" represents "Agree Very Strongly." Responses to the four items are summed so that the resulting scale score might range from 4 to 28, with higher scores representing higher overall approval.

Although subjects complete this process for all six office holders, you are interested only in their evaluations of office holder #4. The profile for this office holder indicates that he is a member of the House of Representatives in the U.S. Congress. It indicates that he has held this office for seven years, and lists some of his accomplishments. The profile then goes on to say that there are reports that the congressman has been implicated in a scandal. The scandal involves using public funds improperly for personal purchases.

Manipulating the predictor (independent) variable: Credibility of the news source. The independent variable in your study is the credibility of the news source that implicates office holder #4 in the scandal. Subjects are randomly assigned to one of three treatment conditions for this independent variable. The profile for office holder #4 is identical in all three treatment conditions with respect to all areas of information except one: the description of the news source that has implicated the congressman in the scandal. This description is manipulated in the following three ways:

- **The low-credibility condition**. For the nine subjects in the low-credibility condition, the profile for office holder #4 indicates that an anonymous news source implicated the congressman in the scandal. You can assume that previous research has established that most subjects view an anonymous news source as a low-credibility news source.
- **The moderate-credibility condition**. For the nine subjects in the moderate-credibility condition, the profile for office holder #4 indicates that the news source was a known political enemy who implicated the congressman in the scandal. Assume that previous research has established that most subjects view a known political enemy as a moderate-credibility news source.
- **The high-credibility condition**. For the nine subjects in the high-credibility condition, the profile for office holder #4 indicates that the news source was an independent investigative agency that implicated the congressman in the scandal. Assume that previous research has established that most subjects view an independent investigative agency as a high-credibility news source.

Research design. The research design for the present study is presented in Figure 15.E2.1:

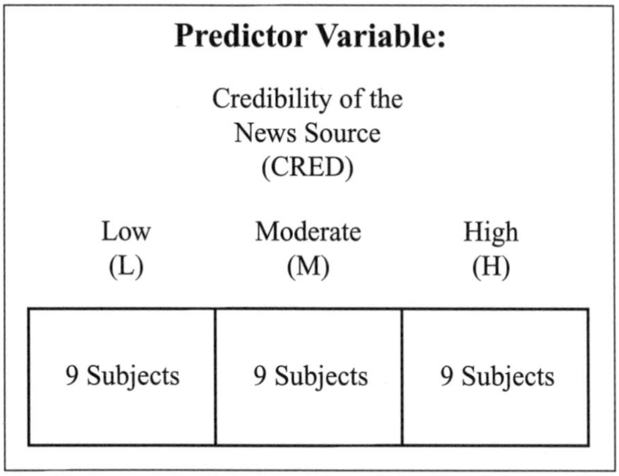

Predictor Variable:
Credibility of the News Source (CRED)
Low (L)
Moderate (M)
High (H)
9 Subjects
9 Subjects
9 Subjects

Figure 15.E2.1: Research Design Used in the Voter Approval Study.

The Predictor and Criterion Variables in the Analysis. The predictor variable in your study is "credibility of the news source." This is an ordinal-level variable that consists of three conditions: (a) low credibility, (b) moderate credibility, and (c) high credibility. In your analysis, you will use the SAS variable name "CRED" to define group membership for this predictor.

The criterion variable in your study is "voter approval:" subject ratings for the overall approval measure described in the previous section of this exercise. You may assume that this variable is on an interval scale, as it has approximately equal intervals, but no true zero point. In your analysis, you will use the SAS variable name "APPROV" to represent this variable.

Data Set to be Analyzed

Table 15.E2.1 presents the data set that you will analyze in this exercise.

Table 15.E2.1

Data from the Voter Approval Study

Subject	Credibility	Voter approval
01	L	18
02	L	21
03	L	20
04	L	20
05	L	24
06	L	22
07	L	19
08	L	16
09	L	19
10	M	20
11	M	22
12	M	21
13	M	15
14	M	24
15	M	19
16	M	18
17	M	19
18	M	17
19	H	14
20	H	18
21	H	16
22	H	19
23	H	15

24	H	16
25	H	11
26	H	12
27	H	14

Each row in Table 15.E2.1 (running horizontally) provides data for a single subject. For example, the first subject has the value "01" for Subject, the value "L" for Credibility, and a score of "18" for Voter approval.

The first column (running vertically) in Table 15.E2.1 is headed "Subject." This column assigns a unique subject number to each participant in the study.

The second column (running vertically) in Table 15.E2.1 is headed "Credibility." This column indicates the group to which each subject was assigned for the predictor variable. The first nine subjects are identified with the value "L," which represents "low credibility." The next nine subjects are identified with the value "M," which represents "moderate credibility." The last nine subjects are identified with the value "H," which represents "high credibility."

The third column in Table 15.E2.1 is headed "Voter approval." This column provides each subject's score for the dependent variable: approval ratings for the overall performance of office holder #4. You will remember that this measure could range from 4 to 28, with higher scores indicating greater approval.

When you write your SAS program, you will type your data as it appears in Table 15.E2.1; you will not need to reorganize the data.

Your Assignment

1. Create a SAS program that will input the data set in Table 15.E2.1. Use PROC GLM to perform a one-way ANOVA with one between-subjects factor. In this analysis, the criterion variable should be "voter approval ratings," and the predictor variable should be "credibility of the news source."

   When you write this program, do the following:

- Use the SAS programs presented in Chapter 15 of the *Student Guide* as models.
- When typing your data, type all of the variables from Table 15.E2.1. You *should* type the subject number variable that appears as the first column in this table. Give this variable the SAS variable name SUB_NUM. In typing this SAS variable name, make sure you use an underscore (_) and not a hyphen (-). When typing numbers such as "01," make sure you type a zero (0) and not the letter O.
- Use the SAS variable name "CRED" to represent the predictor variable (credibility of the news source).
- Use the SAS variable name APPROV to represent subject scores for the criterion variable (voter approval ratings).
- When typing your data, use the values that are presented in Table 15.E2.1. For example, use the values "L," "M," and "H" to define conditions for the predictor variable.
- Include two MEANS statements in your program: one to request means and standard deviations, and one to request the Tukey multiple comparison procedure (request the same options that were requested in Chapter 15 of the *Student Guide*).
- Type your full name in the TITLE1 statement so that it will appear in the output.

2. Submit the program for analysis, and, if necessary, correct it so that it runs without errors.

3. On a clean sheet of paper, prepare an ANOVA summary table that summarizes the results of your analysis. In doing this, follow the directions provided in your *Student Guide*, and use the ANOVA summary tables presented in Chapter 15 of your *Student Guide* as models. Be sure to (a) give your table a table number, (b) give your table a title appropriate for this study, (c) include a *p* value note at the bottom of the table that is appropriate for the level of significance observed in your output, and (d) include the R^2 value. If values in the table contain decimals, report those values to two decimal places.

4. On a separate sheet of paper, prepare a table that provides the differences between means, the confidence intervals for differences between means,

and the results of the Tukey tests. This should be a single table, similar to the table presented in the section titled "Prepare a table that presents the results of the Tukey tests and the confidence intervals" from Chapter 15 of the *Student Guide*. Be sure to (a) give your table a table number, (b) give your table a title appropriate for this study, and (c) include a *p* value note at the bottom of the table that is appropriate for the level of significance observed in your output. If values in the table contain decimals, report those values to three decimal places.

5. Assume that you begin with the following research question and research hypothesis:

   - **Research question**. The purpose of this study is to determine whether there is a significant relationship between (a) the credibility of a news source and (b) voter (subject) approval ratings for a congressman implicated in a scandal.
   - **Research hypothesis**. There will be a negative relationship between news source credibility and voter approval ratings for a congressman implicated in a scandal. Specifically, it is predicted that (a) subjects exposed to the high-credibility condition will display lower approval ratings than subjects exposed to the moderate- or low-credibility conditions, and (b) subjects exposed to the moderate-credibility condition will display lower approval ratings than subjects exposed to the low-credibility condition.

   On a separate sheet of paper, prepare an analysis report to summarize your findings, using the format shown in the section titled "Analysis Report for the Aggression Study" from Chapter 15 of the *Student Guide*. You will have to modify that report so that it is relevant to your analysis.

   Underline each heading in your report. Remember to include information for sections A–N (as shown in Chapter 15 of the *Student Guide*) with a figure (for section O). If values in your report contain decimals, round to two decimal places (except for the *p* value, which should be reported to four decimal places).

6. For section O of step 5, prepare a bar chart that plots the means for the three conditions in your study. In doing this, use the figures presented in Chapter 15 of the *Student Guide* as models.

What You Will Hand In

Hand in the following materials stapled together in this order:

1. A printout of your SAS program (including data), your SAS log, and your SAS output files (your pages of output should be in the correct order).
2. A copy of the ANOVA summary table that you have prepared as a part of your report.
3. A copy of the table that provides the differences between means, the confidence intervals for differences between means, and the results of the Tukey tests. This should be a single table that you have prepared as a part of your report.
4. A copy of your report summarizing the results of the analysis.
5. A figure (bar chart) in which you plot the means for the three conditions in your study. This figure should be included with your report summarizing the results of the analysis.

Hint

If your SAS program ran correctly, your output should resemble the following excerpts from the output of a correctly written program. Remember that your program will produce more output than shown in these excerpts.

```
Source        DF     Type III SS     Mean Square    F Value    Pr > F
 CRED          2     131.5555556      65.7777778      10.18    0.0006
```

```
                          The GLM Procedure

         Level of               ------------APPROV-----------
         CRED          N                Mean              Std Dev

         H             9          15.0000000           2.59807621
         L             9          19.8888889           2.31540733
         M             9          19.4444444           2.69773568
```

Exercises for Chapter 16: Factorial ANOVA with Two Between-Subjects Factors

Exercise 16.1: The Effects of Misleading Suggestions and Pre-Event Instructions On the Creation of False Memories

Overview

In this exercise, you will read about a study that represents an expansion of Exercise 15.1 from the preceding chapter. Exercise 15.1 dealt with a study in which you manipulated a single predictor variable (number of exposures to misleading suggestions), and observed its effect on a criterion variable (confidence that the thief in the video wore gloves—a false memory). For this exercise, you will have two predictor variables. You will analyze the data using factorial analysis of variance (ANOVA) to investigate the nature of the relationship between the two predictor variables and the criterion variable. The following sections describe the study in greater detail.

Note: Although the study reported here is fictitious, it was inspired by the actual study reported by Zaragoza and Mitchell (1996).

The Study

The fictitious study reported in this exercise is very similar to the study in Exercise 15.1, with one difference: the present study includes two predictor variables, instead of only one. The predictor variables will be (a) number of exposures to misleading suggestions (as in Exercise 15.1) and (b) pre-event instructions (the new predictor variable).

In this exercise, you will focus on a research question that deals with the *interaction effect* in the study. Specifically, you will investigate the following research question and hypothesis:

Research question. The purpose of this study is to determine whether there is a significant interaction between (a) the number of exposures to misleading suggestions and (b) pre-event instructions in the prediction of (c) subject confidence that the thief in the video wore gloves.

Research hypothesis. The positive relationship between number of exposures and subject confidence will be stronger for non-warned subjects than for warned subjects.

Research Method

Subjects. Your study includes a total of 24 subjects who are randomly assigned to treatment conditions.

The eyewitness event. Regardless of experimental condition, all subjects in your study view the same 5-minute videotape of a burglary. In the video, two young people burglarize a home and are involved in a car chase with the police.

Manipulating Predictor A: Number of exposures to misleading suggestions. Immediately after viewing the video, each subject completes a 36-item questionnaire. This questionnaire contains 36 YES/NO items such as:

> At the beginning of the video, a blonde-haired youth wearing a leather jacket entered the home. Did he gain entry through a window?

This questionnaire is used to manipulate Predictor A: the number of exposures to misleading suggestions. In this investigation, a **misleading suggestion** is operationalized as a question that presupposes the existence of events or objects that were not actually in the video (although they were plausible). For example, assume that your video does show that the blonde-haired burglar wore a leather jacket, but did not wear gloves. Assume further that you wish to provide a given subject with a misleading suggestion that the burglar *did* wear gloves. To do this, you include the following item in your post-video questionnaire:

> At the beginning of the video, a blonde-haired youth wearing a leather jacket and gloves entered the home. Did he gain entry through a window?

Notice that the preceding question has been modified so that it now presupposes that the burglar was wearing gloves (even though, in reality, he was not). In this way, you have now provided your subject with a misleading suggestion.

You manipulate this independent variable by varying the number of misleading suggestions that appear in the post-video questionnaire. Subjects assigned to the different treatment conditions see different numbers of misleading suggestions:

- The eight subjects assigned to the zero-exposures condition are exposed to no misleading suggestions.
- The eight subjects assigned to the two-exposures condition are exposed to two misleading suggestions.
- The eight subjects assigned to the four-exposures condition are exposed to four misleading suggestions.

Manipulating Predictor B: pre-event instructions. Prior to watching the video, subjects receive different instructions, depending on the treatment condition to which they have been assigned under Predictor B: pre-event instructions:

- The 12 subjects in the warned condition are told "You are about to watch a short videotape depicting a burglary. As you watch the tape, pay particular attention to whether either of the two burglars is wearing gloves. At a later point in time, we will ask you a number of questions about what you saw in the tape."

- The 12 subjects in the non-warned condition are told, "You are about to watch a short videotape depicting a burglary. At a later point in time, we will ask you a number of questions about what you saw in the tape."

You can see that the subjects in the warned condition are told to pay particular attention to whether either of the burglars is wearing gloves, while the subjects in the non-warned condition are not told to pay particular attention to this. The purpose of this manipulation is to determine whether warning subjects protects them from developing false memories about the gloves. If this manipulation is successful, the number-of-exposures independent variable should have an effect on subjects in the non-warned condition, but not on the subjects in the warned condition.

Research design. The present study is a 2 × 3 factorial design, as is illustrated in figure 16.E1.1. You can see that the three levels of Predictor A (number of exposures) are represented as columns running vertically, and the two levels of Predictor B (pre-event instructions) are represented as rows running horizontally.

			Predictor A: Number of Exposures (NUM_EXP)	
		Zero Exposures (0_EXP)	Two Exposures (2_EXP)	Four Exposures (4_EXP)
Predictor B: Pre-Event Instructions (INSTRUCT)	Warned (W)	Cell W-0_EXP 4 Subjects	Cell W-2_EXP 4 Subjects	Cell W-4_EXP 4 Subjects
	Non-Warned (N)	Cell N-0_EXP 4 Subjects	Cell N-2_EXP 4 Subjects	Cell N-4_EXP 4 Subjects

Figure 16.E1.1. Experimental design used in the false memories study that included two predictor variables.

Measuring the criterion variable. The criterion (dependent) variable in your study is the subjects' confidence that the blonde-haired thief wore gloves in the video. As you know, this thief did not wear gloves in the video, so it is reasonable to assume that those subjects who are confident that he did wear gloves are displaying a false memory.

To assess this dependent variable, you ask the subjects to return for another session one week after they watched the video. In this second session, you ask them to complete an 80-item questionnaire in which they indicate whether they believe that certain events occurred in the burglary video that they had previously viewed.

The questionnaire lists a number of possible facts and events, such as:

- The thief stole a lamp.
- The house lights were all off.
- The police car's siren was on.
- The blonde-haired thief wore gloves.

For each item, subjects indicate how confident they are that the video actually portrayed this event. They make their ratings using a 7-point response format in which "7" represents "Definitely Yes" and "1" represents "Definitely No."

For four of the items in the questionnaire, subjects indicate their confidence that "The blonde-haired thief wore gloves" (this idea will be worded somewhat differently in each of the four different items). When you analyze your data, you focus on subject responses to these four items only, and ignore the other 76 items in the questionnaire.

For each subject, you compute his or her average response to the four items dealing with the idea that the thief wore gloves. The resulting mean serves as that subject's score for the dependent variable. For a given subject, this score could range from a low of 1.00 (indicating that subject definitely does not believe that the thief wore gloves), to a high of 7.00 (indicating that subject definitely does believe that the thief wore gloves). Because you are averaging responses to four items, individual scores for this dependent variable could also be fractional values such as 1.75, 3.50, 4.25, and so forth.

The Predictor and Criterion Variables in the Analysis. Predictor A in your study is the "number of exposures to misleading suggestions." This is a limited-value variable, is assessed on a ratio scale, and includes three levels: (a) zero exposures, (b) two exposures, and (c) four exposures. In your analysis, you will use the SAS variable name "NUM_EXP" to define group membership for this predictor.

Predictor B in your study is the "pre-event instructions." This is a dichotomous variable, is assessed on a nominal scale, and incudes two levels: warned and non-warned. In your analysis, you will use the SAS variable name "INSTRUCT" to define group membership for this predictor.

The criterion variable in your study is "confidence that the thief wore gloves:" subject ratings for the measure described in the previous section of this exercise. This is a multi-value variable, and you may assume that it is assessed on an interval scale, as it has equal intervals but no true zero point. In your analysis, you will use the SAS variable name "CONFID" to represent this variable.

Data Set to be Analyzed

Table 16.E1.1 presents the data set that you will analyze in this exercise.

Table 16.E1.1

Data from the False Memories Study with Two Predictor Variables

Subject	Pre-event instructions	Number of exposures	Confidence score
01	W	0_EXP	1.50
02	W	0_EXP	1.25
03	W	0_EXP	2.00
04	W	0_EXP	1.75
05	W	2_EXP	1.50
06	W	2_EXP	2.00
07	W	2_EXP	1.25
08	W	2_EXP	1.50
09	W	4_EXP	1.50
10	W	4_EXP	1.75
11	W	4_EXP	2.00
12	W	4_EXP	1.50
13	N	0_EXP	2.50
14	N	0_EXP	2.50
15	N	0_EXP	2.00
16	N	0_EXP	3.00
17	N	2_EXP	5.00
18	N	2_EXP	5.50
19	N	2_EXP	4.00
20	N	2_EXP	4.75
21	N	4_EXP	5.50
22	N	4_EXP	5.75
23	N	4_EXP	5.75
24	N	4_EXP	5.25

Each row in Table 16.E1.1 (running horizontally) provides data for a single subject. For example, the first subject has the value "01" for Subject, the value "W" for Pre-event instructions, the value "0_EXP" for Number of exposures, and a score of "1.50" for Confidence score.

The first column (running vertically) in Table 16.E1.1 is headed "Subject." This column assigns a subject number to each participant in the study.

The second column (running vertically) in Table 16.E1.1 is headed "Pre-event instructions." This column indicates the group to which each subject is assigned for Predictor B. The first twelve subjects (subjects 1–12) are identified with the value "W," which represents warned. The next twelve subjects (subjects 13–24) are identified with the value "N," which represents non-warned.

The third column (running vertically) in Table 16.E1.1 is headed "Number of exposures." This column indicates the group to which each subject was assigned for Predictor A. The first four subjects are identified with the value "0_EXP," which represents zero exposures. The next four subjects are identified with the value "2_EXP," which represents two exposures. The next four subjects are identified with the value "4_EXP," which represents four exposures. The sequence then repeats itself for the 12 subjects in the non-warned condition.

The fourth column in Table 16.E1.1 is headed "Confidence score." This column provides each subject's score for the criterion variable: his or her confidence that the thief wore gloves. You will remember that this measure could range from a low of 1.00 (meaning that the subject definitely does not believe that the thief wore gloves), to a high of 7.00 (indicating that subject definitely does believe that the thief wore gloves).

When you write your SAS program, you will type your data as it appears in Table 16.E1.1; you will not need to reorganize the data.

Your Assignment

1. Create a SAS program that will input the data set in Table 16.E1.1. Use PROC GLM to perform a factorial ANOVA with two between-subjects factors. In this analysis, the criterion variable should be "confidence that the thief wore gloves." Predictor A should be "number of exposures," and Predictor B should be "pre-event instructions."

   When you write this program, do the following:

   - Use the SAS program presented in Chapter 16 of the *Student Guide* as a model.

- When writing the INPUT statement, use the SAS variable name SUB_NUM to represent subject numbers (the first column of information in Table 16.E1.1). When typing this SAS variable name, make sure that you use an underscore (_) and not a hyphen (-). When typing subject numbers such as "01," make sure that you type a zero (0) and not the letter O.
- When writing the INPUT statement, use the SAS variable names presented in Figure 16.E1.1 of this exercise to represent Predictor A and Predictor B. For example, use the SAS variable name NUM_EXP for Predictor A (number of exposures). Be sure to use an underscore (_) to connect the NUM to the EXP. Do not use a hyphen (-).
- When writing the INPUT statement, use the SAS variable name CONFID to represent subject scores for the criterion variable (confidence that the thief wore gloves).
- When typing your data, use the values presented in Table 16.E1.1 of this exercise to represent the various experimental conditions. For example, use the values "0_EXP," "2_EXP," and "4_EXP" to define conditions for Predictor A.
- Include two MEANS statements in your program (as was done in the example program in Chapter 16 of the *Student Guide*).
- Type your full name in the TITLE1 statement so that it will appear in the output.

2. Submit the program for analysis, and, if necessary, correct it so that it runs without errors.

3. On a clean sheet of paper, prepare an ANOVA summary table that summarizes the results of your analysis. In doing this, follow the directions provided in Chapter 16 of your *Student Guide*, and use the ANOVA summary tables presented in Chapter 16 in your *Student Guide* as models. Be sure to (a) give your table a table number, (b) give your table a title appropriate for this study, (c) include p values in the table, and (d) include R^2 values. If values in the table contain decimals, report those values to two decimal places.

4. For this analysis, it is not necessary for you to report confidence intervals for the differences between the means.

5. Assume that you begin with the following research question and research hypothesis:

   Research question. The purpose of this study was to determine whether there was a significant interaction between (a) the number of exposures to misleading suggestions and (b) pre-event instructions in the prediction of (c) subject confidence that the thief in the video wore gloves.

   Research hypothesis. The positive relationship between number of exposures and subject confidence will be stronger for non-warned subjects than for warned subjects.

   On a separate sheet of paper, prepare an analysis report to summarize your findings. Use the relevant analysis reports (dealing with interaction effects) from Chapter 16 of your *Student Guide* as models. You will have to modify your report so that it is relevant to your analysis.

   Underline each heading in your report. Remember to include information for sections A–N (as shown in chapter 16 of the *Student Guide*) with a figure (for section O). If values in your report contain decimals, round to two decimal places (except for the *p* value, which should be reported to four decimal places).

6. For section O of step 5, prepare a figure that plots the means for the six cells in this research design, similar to the figures that appear in Chapter 16 of your *Student Guide*. In doing this, follow the directions provided in your *Student Guide*, and use the figures presented in your *Student Guide* as models. (Hint: Use a solid line to represent subjects in the non-warned condition for Predictor B, and use a broken line to represent subjects in the warned condition for Predictor B).

What You Will Hand In

Hand in the following materials stapled together in this order:

1. A printout of your SAS program (including data), your SAS log, and your SAS output files (your pages of output should be in the correct order).
2. A copy of the ANOVA summary table that you have prepared as a part of your report.

3. A copy of your report summarizing the results of the analysis (for sections A–N).
4. A figure in which you plot the means for the six cells in your study. This figure (section O) should be included with your report summarizing the results of the analysis.

Hint

If your SAS program ran correctly, your output should resemble the following excerpts from the output of a correctly written program. Remember that your program will produce more output than shown in these excerpts.

Source	DF	Type III SS	Mean Square	F Value	Pr > F
INSTRUCT	1	42.66666667	42.66666667	292.57	<.0001
NUM_EXP	2	10.39583333	5.19791667	35.64	<.0001
INSTRUCT*NUM_EXP	2	10.02083333	5.01041667	34.36	<.0001

Level of INSTRUCT	Level of NUM_EXP	N	------------CONFID----------- Mean	Std Dev
N	0_EXP	4	2.50000000	0.40824829
N	2_EXP	4	4.81250000	0.62500000
N	4_EXP	4	5.56250000	0.23935678
W	0_EXP	4	1.62500000	0.32274861
W	2_EXP	4	1.56250000	0.31457643
W	4_EXP	4	1.68750000	0.23935678

Reference

Zaragoza, M. S., & Mitchell, K. J. (1996). Repeated exposure to suggestion and the creation of false memories. *Psychological Science*, *7*, 294-300.

Exercise 16.2: The Effect of News Source Credibility and Nature of the Scandal on Voter Reactions to Political Scandal

Overview

In this exercise, you will read about a study that represents an expansion of Exercise 15.2 from the preceding chapter. In Exercise 15.2, you analyzed data from a fictitious study in which only one predictor variable (independent variable) was manipulated. In contrast, the study in this exercise will have two predictor variables instead of one only. In this exercise, you will use PROC GLM to perform a factorial ANOVA with two between-subjects factors.

The Study

In Exercise 15.2, you asked subjects to review a profile of a congressman who was said to be involved in a scandal. Subjects provided "voter approval" ratings in which they rated the extent to which they approved of the congressman's overall performance. The predictor variable in your study was the *credibility of the news source:* One-third of the subjects were told that the news source that implicated the congressman in the scandal was a low-credibility news source, one-third were told that it was a moderate-credibility news source, and one-third were told that it was a high-credibility news source. You predicted that there would be a negative relationship between the credibility of the news source and voter approval ratings (in other words, you predicted that, the more credible the news source, the lower the mean voter approval rating would be for the congressman).

For the current exercise, you will replicate the study from Exercise 15.2, and will add a second predictor variable: *the nature of the scandal.* To manipulate this independent variable, half of your subjects will be told that the politician's scandal is personal in nature (in other words, he is accused of personal income tax fraud). The other half will be told that the scandal is public in nature (in other words, he is accused of misusing public funds).

In the present study, your research hypothesis predicts that there will be an interaction between the credibility of the news source and the nature of the scandal. Again, you predict that there will be a negative relationship between news source credibility and voter approval ratings. However, in this study you predict that (a) the relationship between credibility and voter approval will be relatively strong for subjects in the *public* scandal condition, but that (b) the relationship will be relatively weaker for subjects in the *personal* scandal condition.

Research question. The purpose of this study is to determine whether there is a significant interaction between (a) the credibility of the news source and (b) the nature of the scandal in the prediction of (c) voter (subject) approval ratings for a congressman implicated in a scandal.

Research hypothesis. The negative relationship between news source credibility and voter approval ratings will be stronger for subjects in the public funds misuse condition than for subjects in the personal tax fraud condition.

Research Method

Subjects. Your study includes a total of 30 subjects who are randomly assigned to treatment conditions.

Profiles of office-holders. Regardless of the experimental condition, all subjects in your study are asked to read profiles of six politicians holding a variety of offices. Each profile describes the office held, provides background information for the politician (for example, education and work history), describes the politician's achievements while in office, and provides other miscellaneous information.

Measuring the criterion variable. After reading the profile for a given politician, subjects complete a 4-item "voter approval" scale for that office holder. Subjects play the role of voters in the district represented by that office holder, and rate the extent to which they approve of the politician's overall performance. The scale they complete includes items such as "I approve of this office holder's performance." They respond to each item using a 7-point response format in which "1" represents "Disagree Very Strongly" and "7" represents "Agree Very Strongly." Responses to the four items are summed so that the resulting scale score might range from 4 to 28 with higher scores representing higher overall approval.

Although subjects complete this process for all six office holders, you are interested only in their evaluations of Office Holder #4. The profile for this office holder indicates that he is a member of the House of Representatives in the U.S. Congress. It indicates that he has held this office for seven years, and lists some of his accomplishments. The profile then goes on to say that there are reports that the congressman has been implicated in a scandal. It is with this description of a scandal that you manipulate the two predictor variables in your study.

Manipulating Predictor A: the credibility of the news source. The first predictor variable in your study is the credibility of the news source that implicates Office Holder #4 in the scandal. Subjects are randomly assigned to one of three treatment conditions for this independent variable. Different versions of the profile for Office Holder #4 vary the description of the news source that has implicated the congressman in the scandal. This description is manipulated in the following three ways:

- **The low-credibility condition.** For the 10 subjects in the low-credibility condition, the profile for Office Holder #4 indicates that the news source was an anonymous news source that implicated the congressman in the scandal. You can assume that previous research has established that most subjects view an anonymous news source as a low-credibility news source.
- **The moderate-credibility condition.** For the 10 subjects in the moderate-credibility condition, the profile for Office Holder #4 indicates that the news source was a known political enemy who implicated the congressman in the scandal. Assume that previous research has established that most subjects view a known political enemy as a moderate-credibility news source.
- **The high-credibility condition.** For the 10 subjects in the high-credibility condition, the profile for Office Holder #4 indicates that the news source was an independent investigative agency that implicated the congressman in the scandal. Assume that previous research has established that most subjects view an independent investigative agency as a high-credibility news source.

Manipulating Predictor B: the nature of the scandal. The profile for Office Holder #4 is also varied so as to manipulate the study's second independent variable. All subjects are randomly assigned to one of two treatment conditions for the "nature of the scandal" predictor variable:

- **The personal tax fraud condition.** One-half of the subjects are assigned to the personal tax fraud condition. For the 15 subjects in this group, the profile for Office Holder #4 indicates that the congressman might be guilty of tax fraud on his personal income taxes. Assume that prior research shows that most subjects view this as a "personal" scandal.
- **The public funds misuse condition.** One-half of the subjects are assigned to the public funds misuse condition. For the 15 subjects in this group, the profile for Office Holder #4 indicates that the congressman might be guilty of misusing public (government) funds. Assume that prior research shows that most subjects view this as a "public" scandal.

Research design. The present study is a 2×3 factorial design, as is illustrated in Figure 16.E2.1. You can see that the three levels of Predictor A (credibility of the news source) are represented as columns running vertically, and the two levels of Predictor B (nature of the scandal) are represented as rows running horizontally.

		Predictor A: Credibility of the News Source (CRED)		
		Low (L)	Moderate (M)	High (H)
Predictor B: Nature of the Scandal (NATURE)	Personal (PERS)	Cell PERS-L 5 Subjects	Cell PERS-M 5 Subjects	Cell PERS-H 5 Subjects
	Public (PUBLIC)	Cell PUBLIC-L 5 Subjects	Cell PUBLIC-M 5 Subjects	Cell PUBLIC-H 5 Subjects

Figure 16.E2.1: Research design used in the voter approval study.

The Predictor and Criterion Variables in the Analysis. The study involves two predictor variables and one criterion variable. Predictor A is "credibility of the news source." This is a limited-value variable, is assessed on an ordinal scale, and consists of three levels: (a) low credibility, (b) moderate credibility, and (c) high credibility. In your analysis, you will use the SAS variable name "CRED" to define group membership for this predictor.

Predictor B is "nature of the scandal." This is a dichotomous variable, is assessed on a nominal scale, and consists of two levels: (a) a personal tax fraud condition, and, (b) a public funds misuse condition. In your analysis, you will use the SAS variable name "NATURE" to define group membership for this predictor.

Finally, the criterion variable in your study is "voter approval:" subject ratings for the overall approval measure described in the previous section of this exercise. This is a multi-value variable, and you can assume that it is assessed on an interval scale, as it has approximately equal intervals, but no true zero point. In your analysis, you will use the SAS variable name "APPROV" to represent this variable.

Data Set to be Analyzed

Table 16.E2.1 presents the data set that you will analyze in this exercise.

Table 16.E2.1

Data from the Voter Approval Study with Two Predictor Variables

Subject	Nature of scandal	Credibility	Voter approval
01	PERS	L	28
02	PERS	L	23
03	PERS	L	24
04	PERS	L	27
05	PERS	L	20
06	PERS	M	22
07	PERS	M	26
08	PERS	M	18
09	PERS	M	22
10	PERS	M	23
11	PERS	H	20
12	PERS	H	24
13	PERS	H	16
14	PERS	H	19
15	PERS	H	20
16	PUBLIC	L	22
17	PUBLIC	L	26
18	PUBLIC	L	18
19	PUBLIC	L	22
20	PUBLIC	L	23
21	PUBLIC	M	16
22	PUBLIC	M	20
23	PUBLIC	M	12
24	PUBLIC	M	17
25	PUBLIC	M	15
26	PUBLIC	H	10
27	PUBLIC	H	14
28	PUBLIC	H	6
29	PUBLIC	H	11
30	PUBLIC	H	11

Each row in Table 16.E2.1 (running horizontally) provides data for a single subject. For example, the first subject has the value "01" for Subject, the value “PERS” for the heading Nature of scandal, the value “L” for the heading Credibility, and a score of “28” for the heading Voter approval.

The first column (running vertically) in Table 16.E2.1 is headed Subject. This column assigns a unique subject number to each participant in the study.

The second column (running vertically) in Table 16.E2.1 is headed Nature of scandal. This column indicates the group to which each subject was assigned for Predictor B. The first 15 subjects are identified with the value “PERS,” which represents personal scandal. The last 15 subjects are identified with the value “PUBLIC,” which represents public scandal.

The third column in Table 16.E2.1 is headed Credibility. This column indicates the group to which each subject was assigned for Predictor A. The first five subjects are identified with the value “L,” which represents low credibility. The next five subjects are identified with the value “M,” which represents moderate credibility. The next five subjects are identified with the value “H,” which represents high credibility. For the next 15 subjects, this sequence repeats itself.

The fourth column in Table 16.E2.1 is headed Voter approval. This column provides each subject’s score for the dependent variable: approval ratings for the overall performance of Office Holder #4. You will remember that this measure could range from 4 to 28, with higher scores indicating greater approval.

When you write your SAS program, you will type your data as it appears in Table 16.E2.1; you will not need to reorganize the data.

Your Assignment

1. Create a SAS program that will input the data set in Table 16.E2.1. Use PROC GLM to perform a factorial ANOVA with two between-subjects factors. In this analysis, the criterion variable should be “voter approval.” Predictor A should be “credibility of the news source,” and Predictor B should be “the nature of the scandal.”

When you write this program, do the following:

- Use the SAS program presented in Chapter 16 of the *Student Guide* as a model.
- When typing the INPUT statement, use the SAS variable name SUB_NUM to represent subject numbers (the first column of information in Table 16.E2.1). Be sure to use an underscore (_) to connect the "SUB" to the "NUM." Do not use a hyphen (-).
- When typing the INPUT statement, also use the SAS variable names presented in Figure 16.E2.1 of this *Exercise* book for Predictor A and Predictor B. For example, use the SAS variable name CRED for Predictor A (credibility of the news source).
- When writing the INPUT statement, also use the SAS variable name APPROV to represent subject scores for the voter approval criterion variable.
- When typing your data, use the values presented in Table 16.E2.1 to represent the various experimental conditions. For example, use the values "PERS" and "PUBLIC" to define conditions for Predictor B. Also, use the values "L," "M," and "H" to define conditions for Predictor A.
- Include two MEANS statements in your program (as was done in the example program in Chapter 16 of the *Student Guide*).
- Type your full name in the TITLE1 statement so that it will appear in the output.

2. Submit the program for analysis, and, if necessary, correct it so that it runs without errors.

3. On a clean sheet of paper, prepare an ANOVA summary table that summarizes the results of your analysis. In doing this, follow the directions provided in Chapter 16 of your *Student Guide*, and use the ANOVA summary tables presented in Chapter 16 of your *Student Guide* as models. Be sure to (a) give your table a table number, (b) give your table a title appropriate for this study, (c) include p values in the table, and (d) include R^2 values. If values in the table contain decimals, report those values to two decimal places (except for p values, which should be reported to four decimal places).

4. For this analysis, it is not necessary for you to report confidence intervals for the differences between means.

5. Assume that you begin with the following research question and hypothesis:

   Research question. The purpose of this study is to determine whether there is a significant interaction between (a) the credibility of the news source and (b) the nature of the scandal in the prediction of (c) voter (subject) approval ratings for a congressman implicated in a scandal.

   Research hypothesis. The negative relationship between news source credibility and voter approval ratings will be stronger for subjects in the public funds misuse condition than for subjects in the personal tax fraud condition.

   On a separate sheet of paper, prepare an analysis report to summarize your findings. Use the relevant analysis reports (dealing with interaction effects) from Chapter 16 of your *Student Guide* as models. You will have to modify your report so that it is relevant to your analysis.

   Underline each section in your report. Remember to include information for sections A–N (as shown in Chapter 16 of the *Student Guide*) with a figure (for section O). If values in your report contain decimals, round those values to two decimal places (except for the *p* values, which should be reported to four decimal places).

6. For section O of step 5, prepare a figure that plots the means for the six cells in this research design, similar to the figures that appear in Chapter 16 of your *Student Guide*. In doing this, follow the directions provided in Chapter 16 of your *Student Guide*, and use the figures presented in Chapter 16 of your *Student Guide* as models. (Hint: Use a solid line to represent subjects in the personal tax fraud condition for Predictor B, and use a broken line to represent subjects in the public funds misuse condition).

What You Will Hand In

Hand in the following materials stapled together in this order:

1. A printout of your SAS program (including data), your SAS log, and your SAS output files (your pages of output should be in the correct order).
2. A copy of the ANOVA summary table that you have prepared as part of your report.
3. A copy of your report summarizing the results of the analysis (for sections A–N).
4. A figure in which you plot the means for the six cells in your study. This figure (section O) should be included with the report summarizing the results of the analysis.

Hint

If your SAS program ran correctly, your output should resemble the following excerpts from the output of a correctly written program. Remember that your program will produce more output than shown in these excerpts.

Source	DF	Type III SS	Mean Square	F Value	Pr > F
NATURE	1	264.0333333	264.0333333	30.64	<.0001
CRED	2	336.2666667	168.1333333	19.51	<.0001
NATURE*CRED	2	65.0666667	32.5333333	3.78	0.0375

Level of NATURE	Level of CRED	N	-----------APPROV----------- Mean	Std Dev
PERS	H	5	19.8000000	2.86356421
PERS	L	5	24.4000000	3.20936131
PERS	M	5	22.2000000	2.86356421
PUBLIC	H	5	10.4000000	2.88097206
PUBLIC	L	5	22.2000000	2.86356421
PUBLIC	M	5	16.0000000	2.91547595

EXERCISES 17

Exercises for Chapter 17: Chi-Square Test of Independence

Exercise 17.1: The Relationship Between Sex of Children and Marital Disruption

Overview

Is it possible that the sex of the children in a family is related to the likelihood that the parents will separate or divorce? Some researchers have argued that couples may be somewhat less likely to separate if at least some of their children are boys. They argue that fathers often assume a greater role in raising sons than in raising daughters. If this is true, then it is possible that fathers will be more involved in a family if that family includes boys, and this involvement may decrease the likelihood that the couple will separate.

In this exercise you will test this hypothesis. You will use PROC FREQ to perform a chi-square test of independence to determine the relationship between the sex of the children in a family and the likelihood that the parents will remain together.

The Study

Suppose that you are a sociologist studying longitudinal records regarding a sample of families. First, you verify that each family has only two children. Next you identify the sex of the children in each family; each family will be classified as being either a family with two boys, a family with a boy and a girl, or a family with two girls.

For each of the families, you review the records to determine whether the family experienced marital disruption during the 15 years following the birth of their first child. If the parents are still married and together at the end of this time, you classify them as "intact." If they are either separated or divorced, you classify them as "separated."

You then review the data to determine whether there is a difference between families with two boys, families with a boy and a girl, and families with two girls, with respect to their marital status (intact versus separated) at the end of the 15-year period. If there is a difference, this result will tell you that there is a relationship between the sex of the children and subsequent marital status.

Note: Although the study reported here is fictitious, it was inspired by the actual study reported by Morgan, Lye, and Condran (1988). Also note that, although the researchers in that study observed a relationship, it was much weaker than the relationship shown with the present fictitious data set.

Research question. The purpose of this study is to determine whether there is a relationship between (a) the sex of children and (b) marital status (intact versus separated) at the end of a 15-year interval. Specifically, this study is designed to determine whether there is a difference between families with two boys, families with a boy and a girl, and families with two girls with respect to the parents' marital status (intact versus separated) at the end of a 15-year interval.

Research hypothesis. There will be a relationship between the sex of children and marital status such that (a) a larger percentage of families that have two boys will be intact, compared to families with a boy and a girl or families with two girls, and (b) a larger percentage of families that have a boy and a girl will be intact, compared to families with two girls.

Research Method

Subjects. Subjects in your study consist of 147 families. You have verified that each of these families has only two children.

Measuring the predictor variable. The predictor variable in your study is the "sex of the children." This is a limited-value variable, is assessed on a nominal scale of measurement, and consists of three groups:

- families with two boys ($n = 48$),
- families with a boy and a girl ($n = 48$)
- families with two girls ($n = 51$).

In your SAS program you will use the SAS variable name SEX to represent this variable.

Measuring the criterion variable. The criterion variable in your study is the "marital status" of the parents at the end of the 15-year period. This is a dichotomous variable, is assessed on a nominal scale, and consists of two groups: families in which the parents are still intact, and families in which the parents are separated. You will use the SAS variable name STATUS to represent this variable.

The Observational Unit. You will remember that the **observational unit** in a study is the individual subject (or other entity) that serves as the source of the data. In most analyses in this text, the individual *person* has served as the observational unit. In this study, however, the individual *family* serves as the observational unit. This is because the analysis is performed on 147 families, not on 147 individual people. Each family contributed two values to the analysis: (a) the sex of the children in the family, and (b) the marital status of the parents.

Data Set to be Analyzed

In a two-way classification table. Figure 17.E1.1 presents the two-way classification table for your study. The vertical columns of this figure identify the sex of the children in each family. The first column represents the families with two boys, the second column represents the families with a boy and a girl, and the third column represents the families with two girls.

		Sex of the Children in the Family		
		Boy-Boy	Boy-Girl	Girl-Girl
Marital Status	Intact	n=34	n=26	n=15
	Separated	n=14	n=22	n=36

Figure 17.E1.1. Two-way classification table: Marital status of parents as a function of the sex of the children in the family.

The horizontal rows of Figure 17.E1.1 identify the marital status of the parents after 15 years. The top row represents the couples who were still intact, and the bottom row represents the couples who were separated.

Assume that you have already tabulated the number of families that appear in each cell. You can see that

- there were 34 families that (a) included two boys, and (b) were still intact
- there were 26 families that (a) included a boy and a girl, and (b) were still intact.

The remaining cells can be interpreted in the same way.

In a data table. The results presented in Figure 17.E1.1 can also be summarized in the form of a data table. Table 17.E1.1 shows the same frequencies, this time rearranged so that they can be analyzed using PROC FREQ.

Table 17.E1.1

Tabular Data Set for the Family Disruption Study

Marital status[a]	Sex of children[b]	Number
I	BB	34
I	BG	26
I	GG	15
S	BB	14
S	BG	22
S	GG	36

[a] For marital status, "I" = intact and "S" =separated.
[b] For sex of children, "BB" = families with two boys, "BG" = families with a boy and a girl, and "GG" = families with two girls.

The first column of Table 17.E1.1 defines marital status. The value "I" represents the intact families, and the value "S" represents the separated families. The second column defines the sex of the children in the family: "BB" represents families with two boys, "BG" represents families with a boy and a girl, and "GG" represents families with two girls. The third column of the table indicates how many families there were in each cell (from Figure 17.E1.1).

Each row in Table 17.E1.1 (running horizontally) provides data for a single cell from Figure 17.E1.1. The first row represents the families in which (a) the parents remained intact and (b) the children consisted of two boys. For Number, you can see that there were 34 families in this cell. The remaining rows may be interpreted in the same way.

To make your assignment a bit easier, when you prepare your SAS program, you will type your data as it appears in Table 17.E1.1; you do not need to reorganize the data.

Your Assignment

1. Create a SAS program that will input the data set presented in Table 17.E1.1. Use PROC FREQ to perform a chi-square test of independence.

In this analysis, the criterion variable should be marital status and the predictor variable should be sex of children.

When you write this program, do the following:

- Use the SAS program presented in Chapter 17 of the *Student Guide* as a model. When writing the DATA step, make sure that you use the instructions for inputting tabular data, and not the instructions for inputting raw data.
- When writing the INPUT statement, use the SAS variable name STATUS to represent marital status, the criterion variable in your study. Use the SAS variable name SEX to represent the sex of children, the predictor variable in your study. Finally, use the SAS variable name NUMBER to represent the number of families in each cell.
- When typing your data, use the same values that appear in Table 17.E1.1. For example, use the values "I" and "S" to define groups for the criterion variable, STATUS.
- In your TABLES statement, be sure to include the key word "ALL" to request all statistics.
- Add a TITLE1 statement to the end of your program, immediately before the RUN statement. Your TITLE1 statement should resemble the following:

  TITLE1 *'your-full-name'*;

2. Submit the program for analysis, and, if necessary, correct it so that it runs without errors.

3. Assume that you begin with the following research question and research hypothesis:

   Research question. The purpose of this study was to determine whether there was a relationship between (a) the sex of children and (b) marital status (intact versus separated) at the end of a 15-year interval. Specifically, this study was designed to determine whether there was a difference between families with two boys, families with a boy and a girl, and families with two girls with respect to the parents' marital status (intact versus separated) at the end of a 15-year interval.

Research hypothesis. There will be a relationship between the sex of children and marital status such that (a) a larger percentage of families that have two boys will be intact, compared to families with a boy and a girl or families with two girls, and (b) a larger percentage of families that have a boy and a girl will be intact, compared to families with two girls.

On a separate sheet of paper, prepare an analysis report to summarize your findings. Your analysis report should be based on the research question and research hypothesis presented in this exercise. Use the relevant analysis report from the *Student Guide* as a model. You will have to modify that report so that it is relevant to the current study.

Underline each section of your report. Remember to include information for sections A–L (as shown in Chapter 17 of the *Student Guide*) and include your bar chart figure for section M (described in step 4) as the last page of your report.

With respect to rounding values and the number of decimal places for the values that appear in your report:

- Round the chi-square statistic to three decimal places.
- Report the p value for the chi-square statistic to four decimal places.
- Round Cramer's V or the phi coefficient (whichever you use) to two decimal places.
- When you report percentages in your "formal description" section, report them as two-digit numbers.

4. Along with your analysis report for step 3, prepare a bar chart (for section M) that illustrates the frequencies for each cell. When creating the bar chart, use the figures presented in Chapter 17 of the *Student Guide* as models—copy the format of these bar charts as closely as possible. Include a legend (a key) in your figure that indicates what the solid bars represent and what the white bars represent (as was done with the bar charts in the *Student Guide*). Hint: The label for the horizontal axis of this figure should be "Sex of the Children in the Family."

What You Will Hand In

Hand in the following materials stapled together in this order:

1. A printout of your SAS program (including data), your SAS log, and your SAS output files (your pages of output should be in the correct order).
2. A copy of your report summarizing the results of the analysis (for sections A–L).
3. A figure (bar chart) in which you plot frequencies for the six cells in your study. This figure (section M) should be included with your report summarizing the results of the analysis.

Hint

If your SAS program ran correctly, your output should resemble the following excerpt from the output of a correctly written program. Remember that your program will produce more output than shown in this excerpt.

```
                    JOHN DOE                                1

                The FREQ Procedure

              Table of STATUS by SEX

      STATUS      SEX

      Frequency|
      Percent  |
      Row Pct  |
      Col Pct  |BB      |BG      |GG      |  Total
      ---------+--------+--------+--------+
      I        |     34 |     26 |     15 |     75
               |  23.13 |  17.69 |  10.20 |  51.02
               |  45.33 |  34.67 |  20.00 |
               |  70.83 |  54.17 |  29.41 |
      ---------+--------+--------+--------+
      S        |     14 |     22 |     36 |     72
               |   9.52 |  14.97 |  24.49 |  48.98
               |  19.44 |  30.56 |  50.00 |
               |  29.17 |  45.83 |  70.59 |
      ---------+--------+--------+--------+
     Total           48       48       51      147
                  32.65    32.65    34.69   100.00
```

Reference

Morgan, S.P., Lye, D.N., and Condran, G.A. (1988). Sons, daughters, and the risk of marital disruption. *American Journal of Sociology*, *94*, 110-129.

Exercise 17.2: The Relationship Between Membership in College Student Organizations and Sexually Coercive Behavior in Men

Overview

Sexually coercive behavior occurs when one individual attempts to engage in some type of unwanted sexual behavior with another individual. These behaviors may range from the less serious (an unwanted kiss) to the more serious (the use of violence to commit sexual assault).

In recent decades, universities have shown increased interest in decreasing the incidence of sexual assault on campus. One approach to dealing with this problem might involve identifying the type of student who is more likely to engage in sexually coercive behavior. The present fictitious study is an example of this approach.

In this example you will use PROC FREQ to perform a chi-square test of independence to determine the relationship between two variables.

The Study

Supose that you are a sociologist who obtains completed questionnaires from approximately 500 male college students. The questionnaire assesses (a) whether the student has ever engaged in an act of sexual coercion since enrolling in college, and (b) the type of student organization to which the student belongs (if any). With respect to the second variable, you classify each student as either

- not belonging to any student organization
- belonging to a social fraternity
- belonging to some type of student organization other than a social fraternity (for example, student government organization).

Your research hypothesis is that a larger percentage of men who belong to social fraternities will have engaged in sexually coercive behavior since arriving on campus, compared to the men in the other two groups. You will use a chi-square test of independence to test this hypothesis.

Note: Although the study reported here is fictitious, it was inspired by the actual study reported by Garrett-Gooding and Senter (1987).

Research question. The purpose of this study was to determine whether there was a relationship between (a) membership in college student organizations and (b) engagement in sexually coercive behavior in a sample of male college students. Specifically, this study was designed to determine whether there was a difference between men who belong to no organizations, men who belong to social fraternities, and men who belong to non-fraternity organizations with respect to their engagement in sexually coercive behavior.

Research hypothesis. There will be a relationship between membership in college student organizations and engagement in sexually coercive behavior such that a larger percentage of men who belong to social fraternities will engage in sexually coercive behavior, compared to men who belong to non-fraternity organizations or men who belong to no organizations.

Research Method

Subjects. Subjects in your study consist of 483 men enrolled at a medium-size university.

Measuring the predictor variable. The predictor variable in your study is "type of organization." This is a limited-value variable, is assessed on a nominal-level scale of measurement, and consists of three groups:

- men who do not belong to any student organization ($n = 211$)
- men who belong to a social fraternity ($n = 75$)
- men who belong to a student organization other than a social fraternity ($n = 197$).

You will use the SAS variable name TYPE_ORG to represent this variable.

Measuring the criterion variable. The criterion variable in your study is the men's "engagement in sexually coercive behavior." Assume that you administer a questionnaire to the subjects that lists 15 sexually coercive acts that range from the less serious (such as "Kissed someone who does not want to be kissed") to the more serious (such as "Used violence to obtain sexual intercourse"). For each item, subjects indicate the number of times that they have engaged in that behavior since enrolling in college (for each item the respondent *can* indicate "Never").

Based on their responses to the questionnaire, you divide the respondents into two groups. The "never" group (n = 265) consists of the subjects who indicate "Never" to all 15 of the sexually coercive acts. The "some" group (n = 218) consists of the subjects who indicate that they have committed at least one of the sexually coercive acts since enrolling in college.

The criterion variable in your study is therefore "engagement in sexually coercive behavior." This variable is a dichotomous variable, because it can assume only two values: "never" versus "some." Specifically, the subjects in the study are divided into two groups for this variable:

- men who indicate that they have "never" engaged in sexually coercive behavior (n = 265)
- men who indicate that they have engaged in "some" sexually coercive behavior (n = 218).

This criterion variable is assessed on an ordinal scale, since it provides a primitive hierarchy of levels with respect to the construct of "sexually coercive behavior" (in other words, the subjects in the "some" group have presumably engaged in this behavior more than the subjects in the "never" group). You will use the SAS variable name COERC to represent this criterion variable.

Data Set to be Analyzed

In a two-way classification table. Figure 17.E2.1 presents the two-way classification table for your study. The vertical columns of this figure identify the type of organization to which a subject belongs. The first column represents the men who do not belong to any student organizations, the second represents the men who belong to a social fraternity, and the third column represents the men who belong to a student organization other than a social fraternity.

		Type of Organization		
		None	Fraternity	Other
Engagement in Sexually Coercive Behavior	Never	n=120	n=27	n=118
	Some	n=91	n=48	n=79

Figure 17.E2.1. Two-way classification table: Men's engagement in sexually coercive behavior as a function of membership in student organizations.

The horizontal rows of Figure 17.E2.1 identify the levels of the "engagement in sexually coercive behavior" criterion variable. The top row represents the men who indicate that they have never engaged in sexually coercive behavior, and the bottom row represents the men who indicate that they have engaged in some sexually coercive behavior.

Assume that you have already tabulated the number of subjects that appear in each cell of Figure 17.E2.1. You can see that

- there were 120 men in the cell for the subjects who (a) do not belong to any student organizations ("None"), and (b) indicate that they have never engaged in sexually coercive behavior ("Never")
- there were 27 men in the cell for the subjects who (a) belong to a social fraternity ("Fraternity"), and (b) indicate that they have never engaged in sexually coercive behavior ("Never").

The remaining cells can be interpreted in the same way.

In a data table. The results presented in Figure 17.E2.1 can also be summarized in the form of a data table. Table 17.E2.1 shows the same frequencies, this time rearranged so that they can be analyzed using PROC FREQ.

Table 17.E2.1

Data from the Sexually Coercive Behavior Study

Coercive behavior	Type of organization[a]	Number
NEVER	NONE	120
NEVER	FRAT	27
NEVER	OTHER	118
SOME	NONE	91
SOME	FRAT	48
SOME	OTHER	79

[a] For "Type of organization," "NONE" = men who do not belong to any student organization, "FRAT" = men who belong to social fraternities, and "OTHER" = men who belong to student organizations other than social fraternities.

The first column of Table 17.E2.1 defines engagement in sexually coercive behavior. The value "NEVER" represents men who indicate that they have never engaged in sexually coercive behavior, and the value "SOME" represents men who indicate that they have engaged in some sexually coercive behavior.

The second column defines the type of organization to which the subject belongs: "NONE" represents men who do not belong to any student organizations, "FRAT" represents men who belong to social fraternities, and "OTHER" represents men who belong to student organizations other than social fraternities.

The third column of the table indicates how many subjects appeared in that cell.

Each row in Table 17.E2.1 (running horizontally) provides data for a single cell from Figure 17.E2.1.The first row represents the subjects who indicate that (a) they have never engaged in sexually coercive behavior, and (b) they do not belong to any student organization. For Number, you can see that there were 120 subjects in this cell. The remaining rows can be interpreted in the same way.

When you write your SAS program, you will type your data as it appears in Table 17.E2.1; you do not need to reorganize the data.

Your Assignment

1. Create a SAS program that will input the data set presented in Table 17.E2.1. Use PROC FREQ to perform a chi-square test of independence. In this analysis, the criterion variable should be engagement in sexually coercive behavior ("never" versus "some") and the predictor variable should be type of organization ("none" versus "fraternity" versus "other").

When you write this program, do the following:

- Use the SAS program presented in Chapter 17 of the *Student Guide* as a model. When writing the DATA step, make sure that you use the instructions for inputting tabular data, and not the instructions for inputting raw data.
- When writing the INPUT statement, use the SAS variable name COERC to represent engagement in sexually coercive behavior, the criterion variable in your study.
- Use the SAS variable name TYPE_ORG to represent the type of organization, the predictor variable in your study. When typing this SAS variable name, be sure to use an underscore (_), and not a hyphen (-).
- Finally, use the SAS variable name NUMBER to represent the number of subjects in each cell.
- When typing your data, use the same values that appear in Table 17.E2.1. For example, use the values "NEVER" and "SOME" to define groups for the criterion variable, COERC.
- In your TABLES statement, be sure to include the keyword "ALL" to request all statistics.

- Add a TITLE1 statement to the end of your program, immediately before the RUN statement. Your TITLE1 statement should resemble the following:

  TITLE1 *'your- full- name'*;

2. Submit the program for analysis, and, if necessary, correct it so that it runs without errors.

3. Assume that you begin with the following research question and research hypothesis:

   Research question. The purpose of this study was to determine whether there was a relationship between (a) membership in college student organizations and (b) engagement in sexually coercive behavior in a sample of male college students. Specifically, this study was designed to determine whether there was a difference between men who belong to no organizations, men who belong to social fraternities, and men who belong to non-fraternity organizations with respect to their engagement in sexually coercive behavior.

   Research hypothesis. There will be a relationship between membership in college student organizations and engagement in sexually coercive behavior such that a larger percentage of men who belong to social fraternities will engage in sexually coercive behavior, compared to men who belong to non-fraternity organizations or men who belong to no organizations.

   On a separate sheet of paper, prepare an analysis report to summarize your findings. Your analysis report should be based on the research question and research hypothesis presented in this exercise. Use the relevant analysis report from the *Student Guide* as a model. You will have to modify that report so that it is relevant to the current study.

   Underline each section of your report. Remember to include information for sections A–L (as shown in Chapter 17 of the *Student Guide*) and include your bar chart figure for section M (described in step 4) as the last page of your report.

   With respect to rounding values and the number of decimal places for the values that appear in your report:

   - Round the chi-square statistic to three decimal places.

- Report the *p* value for the chi-square statistic to four decimal places.
- Round Cramer's *V* or the phi coefficient (whichever you use) to two decimal places.
- When you report percentages in your "formal description" section, report them as two-digit numbers (such as "71 percent").

4. Along with your analysis report for step 3, prepare a bar chart (for section M) that illustrates the frequencies for each cell. In doing this, use the figures presented in Chapter 17 of the *Student Guide* as models—copy the format of these bar charts as closely as possible. Include a legend (a key) in your figure that indicates what the solid bars represent and what the white bars represent (as was done with the bar charts in the *Student Guide*). Hint: The label for the horizontal axis of this figure should be "Type of Organization."

What You Will Hand In

Hand in the following materials stapled together in this order:

1. A printout of your SAS program (including data), your SAS log, and your SAS output files (your pages of output should be in the correct order).
2. A copy of your report summarizing the results of the analysis (for sections A–L).
3. A figure (bar chart) in which you plot frequencies for the six cells in your study. This figure (section M) should be included with your report summarizing the results of the analysis.

Hint

If your SAS program ran correctly, your output should resemble the following excerpt from the output of a correctly written program. Remember that your program will produce more output than shown in this excerpt.

```
                     JANE DOE                          1

                The FREQ Procedure

           Table of COERC by TYPE_ORG

       COERC      TYPE_ORG

       Frequency|
       Percent  |
       Row Pct  |
       Col Pct  |FRAT    |NONE    |OTHER   |  Total
       ---------+--------+--------+--------+
       NEVER    |     27 |    120 |    118 |    265
                |   5.59 |  24.84 |  24.43 |  54.87
                |  10.19 |  45.28 |  44.53 |
                |  36.00 |  56.87 |  59.90 |
       ---------+--------+--------+--------+
       SOME     |     48 |     91 |     79 |    218
                |   9.94 |  18.84 |  16.36 |  45.13
                |  22.02 |  41.74 |  36.24 |
                |  64.00 |  43.13 |  40.10 |
       ---------+--------+--------+--------+
       Total          75      211      197      483
                   15.53    43.69    40.79   100.00
```

Reference

Garrett-Gooding, J. & Senter, R. (1987). Attitudes and acts of sexual aggression on a university campus. *Sociological Inquiry, 57*, 348-371.

SOLUTIONS

Overview

This section contains solutions for the odd-numbered exercises in each chapter, beginning with Chapter 3, "Writing and Submitting SAS Programs." Each solution shows the SAS program used to complete the exercise, the SAS log, the SAS output, and answers to any questions asked in the exercise.

Solution for Chapter 3: Writing and Submitting SAS Programs

3
SOLUTION

Solution to Exercise 3.1: Computing Mean Height, Weight, and Age

The SAS Program

```
OPTIONS  LS=80  PS=60;
DATA D1;
   INPUT  SUB_NUM
          HEIGHT
          WEIGHT
          AGE ;
DATALINES;
1 64 140 20
2 68 170 28
3 74 210 20
4 60 110 32
5 64 130 22
6 68 170 23
7 65 140 22
8 65 140 22
9 68 160 22
;
PROC MEANS DATA=D1;
   VAR  HEIGHT  WEIGHT  AGE;
   TITLE1 'JANE DOE';
RUN;
```

The SAS Log

The log that is produced by your SAS program might differ from the following log. For example, your log file might contain a number of additional lines at the beginning. In addition, the line numbers on the left side of your log might be higher than the line numbers in the following log (for example, your line numbers might begin with 22, 23, 24, and so on).

```
1     OPTIONS  LS=80  PS=60;
2     DATA D1;
3        INPUT  SUB_NUM
4               HEIGHT
5               WEIGHT
6               AGE ;
7     DATALINES;

NOTE: The data set WORK.D1 has 9 observations and 4 variables.
NOTE: DATA statement used:
      real time           1.54 seconds


17    ;
18    PROC MEANS DATA=D1;
19       VAR  HEIGHT  WEIGHT  AGE;
20       TITLE1 'JANE DOE';
21    RUN;

NOTE: There were 9 observations read from the dataset WORK.D1.
NOTE: PROCEDURE MEANS used:
      real time           1.81 seconds
```

The SAS Output

```
                            JANE DOE                              1

                      The MEANS Procedure

Variable  N          Mean       Std Dev       Minimum       Maximum
---------------------------------------------------------------
HEIGHT    9    66.2222222     3.8980052    60.0000000    74.0000000
WEIGHT    9   152.2222222    29.0593263   110.0000000   210.0000000
AGE       9    23.4444444     3.9721251    20.0000000    32.0000000
---------------------------------------------------------------
```

4 SOLUTION

Solution for Chapter 4: Data Input

Solution to Exercise 4.1: Creating a Data Set Containing LAT Test Scores

The SAS Program

```
OPTIONS  LS=80  PS=60;
DATA D1;
   INPUT   SUB_NUM
           LAT_V
           LAT_M
           SEX  $
           TEST_1
           TEST_2
           TEST_3 ;
DATALINES;
01 510 520 F 89 92 92
02 530   . M 88 75 89
03 620 600 F 95 90 88
04   . 490 F 80  . 78
05 650 600 M 97 95 96
06 550 510 F 76 70 78
07 420 480 . 88 81 85
08 400 410 M 90 88 88
09 590 610 F 90 92 95
;
PROC MEANS DATA=D1;
```

Continued on next page

Continued from previous page

```
   VAR  LAT_V  LAT_M  TEST_1  TEST_2  TEST_3;
   TITLE1 'JOHN DOE';
RUN;
PROC FREQ DATA=D1;
   TABLES SEX;
RUN;
PROC PRINT DATA=D1;
RUN;
```

The SAS Log

If your program ran correctly, your SAS log should resemble the following log. If you submitted your program more than once, you might not see all the notes that appear at the top of this log, and your line numbers on the left side might be different.

```
NOTE: SAS initialization used:
      real time           27.34 seconds

1     OPTIONS  LS=80  PS=60;
2     DATA D1;
3        INPUT   SUB_NUM
4                LAT_V
5                LAT_M
6                SEX  $
7                TEST_1
8                TEST_2
9                TEST_3 ;
10    DATALINES;

NOTE: The data set WORK.D1 has 9 observations and 7 variables.
NOTE: DATA statement used:
      real time           1.53 seconds

20    ;
21    PROC MEANS DATA=D1;
22       VAR  LAT_V  LAT_M  TEST_1  TEST_2  TEST_3;
23       TITLE1 'JOHN DOE';
24    RUN;
```

```
NOTE: There were 9 observations read from the data set WORK.D1.
NOTE: PROCEDURE MEANS used:
      real time           1.87 seconds

     PROC FREQ DATA=D1;
        TABLES SEX;
     RUN;

NOTE: There were 9 observations read from the data set WORK.D1.
NOTE: PROCEDURE FREQ used:
      real time           0.66 seconds


     PROC PRINT DATA=D1;
     RUN;

NOTE: There were 9 observations read from the data set WORK.D1.
NOTE: PROCEDURE PRINT used:
      real time           0.17 seconds
```

The SAS Output

```
                               JOHN DOE                              1

                           The MEANS Procedure

Variable  N          Mean       Std Dev        Minimum        Maximum
----------------------------------------------------------------------
LAT_V     8   533.7500000    89.2728562    400.0000000    650.0000000
LAT_M     8   527.5000000    70.8620390    410.0000000    610.0000000
TEST_1    9    88.1111111     6.5849154     76.0000000     97.0000000
TEST_2    8    85.3750000     9.0386077     70.0000000     95.0000000
TEST_3    9    87.6666667     6.5000000     78.0000000     96.0000000
----------------------------------------------------------------------
```

```
                              JOHN DOE                                 2

                         The FREQ Procedure

                                            Cumulative    Cumulative
  SEX    Frequency       Percent           Frequency       Percent
-------------------------------------------------------------------
  F              5         62.50                  5          62.50
  M              3         37.50                  8         100.00

                       Frequency Missing = 1
```

```
                              JOHN DOE                                 3

 Obs  SUB_NUM  LAT_V  LAT_M  SEX  TEST_1  TEST_2  TEST_3

  1      1      510    520    F     89      92      92
  2      2      530      .    M     88      75      89
  3      3      620    600    F     95      90      88
  4      4        .    490    F     80       .      78
  5      5      650    600    M     97      95      96
  6      6      550    510    F     76      70      78
  7      7      420    480          88      81      85
  8      8      400    410    M     90      88      88
  9      9      590    610    F     90      92      95
```

5
SOLUTION

Solution for Chapter 5: Creating Frequency Tables

Solution to Exercise 5.1: Using PROC FREQ to Analyze LAT Data

The SAS Program

```
OPTIONS  LS=80  PS=60;
DATA D1;
   INPUT   SUB_NUM
           LAT_V
           LAT_M
           LAT_A
           MAJOR  $ ;
DATALINES;
01    540    540    540    A
02    510    560    550    A
03    500    520    530    A
04    490    550    530    A
05    520    510    510    A
06    520    500    530    A
07    510    470    520    E
08    530    490    520    B
09    500    460    480    B
10    490    480    500    E
11    510    470    470    B
12    500    450    490    E
13    500    460    540    E
```
Continued on next page

Continued from previous page

```
14    480    440    490    B
15    490    430    460    B
16    480    450    480    E
17    470    440    470    E
18    460    450    480    B
;
PROC FREQ DATA=D1;
   TABLES  LAT_V;
   TITLE1 'JOHN DOE';
   RUN;
```

The SAS Log

```
NOTE: SAS initialization used:
      real time           18.45 seconds

1     OPTIONS  LS=80  PS=60;
2     DATA D1;
3        INPUT   SUB_NUM
4                LAT_V
5                LAT_M
6                LAT_A
7                MAJOR  $ ;
8     DATALINES;

NOTE: The data set WORK.D1 has 18 observations and 5 variables.
NOTE: DATA statement used:
      real time           1.53 seconds



27    ;
28    PROC FREQ DATA=D1;
29       TABLES  LAT_V;
30       TITLE1 'JOHN DOE';
31    RUN;

NOTE: There were 18 observations read from the data set WORK.D1.
NOTE: PROCEDURE FREQ used:
      real time           1.70 seconds
```

The SAS Output (with Answers Circled)

```
                              JOHN DOE                                1

                         The FREQ Procedure

                                                Cumulative    Cumulative
   LAT_V     Frequency      Percent     Frequency       Percent
   -----------------------------------------------------------------
Q1→ (460)            1          5.56             1          5.56
    470              1          5.56             2         11.11
    480              2         11.11       Q5→ (4)         22.22
    490     Q3→    (3)         16.67             7         38.89
    500              4   Q4→ (22.22)            11         61.11
    510              3         16.67            14         77.78
    520              2         11.11            16   Q6→ (88.89)
    530              1          5.56            17         94.44
Q2→ (540)            1          5.56       Q7→ (18)       100.00
```

Exercise 5.1: Answers to Questions

1. **Question**: What is the lowest observed value for the LAT_V variable?

   Answer: 460

2. **Question**: What is the highest observed value for the LAT_V variable?

   Answer: 540

3. **Question**: How many people had an LAT Verbal score of 490? (In other words, what is the frequency for people who displayed a score of 490 on the LAT_V variable?)

   Answer: Three

4. **Question**: What percentage of people had an LAT Verbal score of 500?

   Answer: 22.2%

5. **Question**: How many people had an LAT Verbal score of 480 or lower?

   Answer: Four

6. **Question**: What percentage of people had an LAT Verbal score of 520 or lower?

   Answer: 88.9%

7. **Question**: What is the total number of usable observations for the LAT_V variable in this data set?

   Answer: 18

6 SOLUTION

Solution for Chapter 6: Creating Graphs

Solution to Exercise 6.1: Using PROC CHART to Create Bar Charts from LAT Data

The SAS Program

```
OPTIONS  LS=80  PS=60;
DATA D1;
   INPUT    SUB_NUM
            LAT_V
            LAT_M
            LAT_A
            MAJOR  $ ;
DATALINES;
01    540    540    540    A
02    510    560    550    A
03    500    520    530    A
04    490    550    530    A
05    520    510    510    A
06    520    500    530    A
07    510    470    520    E
08    530    490    520    B
09    500    460    480    B
10    490    480    500    E
11    510    470    470    B
12    500    450    490    E
13    500    460    540    E
14    480    440    490    B
```

Continued on next page

Continued from previous page

```
15   490   430   460   B
16   480   450   480   E
17   470   440   470   E
18   460   450   480   B
;

PROC CHART  DATA=D1;
   VBAR  LAT_V  /  DISCRETE;
   TITLE1  'JOHN DOE';
RUN;

PROC CHART  DATA=D1;
   VBAR  MAJOR  /  SUMVAR=LAT_V  TYPE=MEAN;
   TITLE1  'JOHN DOE';
RUN;
```

The SAS Log

```
NOTE: SAS initialization used:
      real time           26.30 seconds

1     OPTIONS  LS=80  PS=60;
2     DATA D1;
3        INPUT   SUB_NUM
4                LAT_V
5                LAT_M
6                LAT_A
7                MAJOR  $ ;
8     DATALINES;

NOTE: The data set WORK.D1 has 18 observations and 5 variables.
NOTE: DATA statement used:
      real time           1.59 seconds

27    ;
28
29    PROC CHART  DATA=D1;
30       VBAR  LAT_V  /  DISCRETE;
31       TITLE1  'JOHN DOE';
32    RUN;

NOTE: There were 18 observations read from the data set WORK.D1.
NOTE: PROCEDURE CHART used:
      real time           0.93 seconds

33
34    PROC CHART  DATA=D1;
35       VBAR  MAJOR  /  SUMVAR=LAT_V  TYPE=MEAN;
36       TITLE1  'JOHN DOE';
37    RUN;

NOTE: There were 18 observations read from the data set WORK.D1.
NOTE: PROCEDURE CHART used:
      real time           0.17 seconds
```

The SAS Output

```
                                JOHN DOE                          1

Frequency

4 +                                          *****
  |                                          *****
  |                                          *****
  |                                          *****
  |                                          *****
  |                                          *****
  |                                          *****
  |                                          *****
  |                                          *****
  |                                          *****
3 +                                *****     *****     *****
  |                                *****     *****     *****
  |                                *****     *****     *****
  |                                *****     *****     *****
  |                                *****     *****     *****
  |                                *****     *****     *****
  |                                *****     *****     *****
  |                                *****     *****     *****
  |                                *****     *****     *****
  |                                *****     *****     *****
2 +                      *****     *****     *****     *****     *****
  |                      *****     *****     *****     *****     *****
  |                      *****     *****     *****     *****     *****
  |                      *****     *****     *****     *****     *****
  |                      *****     *****     *****     *****     *****
  |                      *****     *****     *****     *****     *****
  |                      *****     *****     *****     *****     *****
  |                      *****     *****     *****     *****     *****
  |                      *****     *****     *****     *****     *****
  |                      *****     *****     *****     *****     *****
1 +  *****     *****     *****     *****     *****     *****     *****     *****     *****
  |  *****     *****     *****     *****     *****     *****     *****     *****     *****
  |  *****     *****     *****     *****     *****     *****     *****     *****     *****
  |  *****     *****     *****     *****     *****     *****     *****     *****     *****
  |  *****     *****     *****     *****     *****     *****     *****     *****     *****
  |  *****     *****     *****     *****     *****     *****     *****     *****     *****
  |  *****     *****     *****     *****     *****     *****     *****     *****     *****
  |  *****     *****     *****     *****     *****     *****     *****     *****     *****
  |  *****     *****     *****     *****     *****     *****     *****     *****     *****
  |  *****     *****     *****     *****     *****     *****     *****     *****     *****
 -----------------------------------------------------------------------------------------------
      460       470       480       490       500       510       520       530       540

                                          LAT_V
```

```
                    JOHN DOE                                          2

           LAT_V Mean

               |          *****
           500 +          *****        *****
               |          *****        *****        *****
               |          *****        *****        *****
               |          *****        *****        *****
               |          *****        *****        *****
           450 +          *****        *****        *****
               |          *****        *****        *****
               |          *****        *****        *****
               |          *****        *****        *****
               |          *****        *****        *****
           400 +          *****        *****        *****
               |          *****        *****        *****
               |          *****        *****        *****
               |          *****        *****        *****
               |          *****        *****        *****
           350 +          *****        *****        *****
               |          *****        *****        *****
               |          *****        *****        *****
               |          *****        *****        *****
               |          *****        *****        *****
           300 +          *****        *****        *****
               |          *****        *****        *****
               |          *****        *****        *****
               |          *****        *****        *****
               |          *****        *****        *****
           250 +          *****        *****        *****
               |          *****        *****        *****
               |          *****        *****        *****
               |          *****        *****        *****
               |          *****        *****        *****
           200 +          *****        *****        *****
               |          *****        *****        *****
               |          *****        *****        *****
               |          *****        *****        *****
               |          *****        *****        *****
           150 +          *****        *****        *****
               |          *****        *****        *****
               |          *****        *****        *****
               |          *****        *****        *****
               |          *****        *****        *****
           100 +          *****        *****        *****
               |          *****        *****        *****
               |          *****        *****        *****
               |          *****        *****        *****
               |          *****        *****        *****
            50 +          *****        *****        *****
               |          *****        *****        *****
               |          *****        *****        *****
               |          *****        *****        *****
               |          *****        *****        *****
               ---------------------------------------------------------
                            A            B            E

                                       MAJOR
```

Majors with the highest mean score: Arts & Science Majors → A

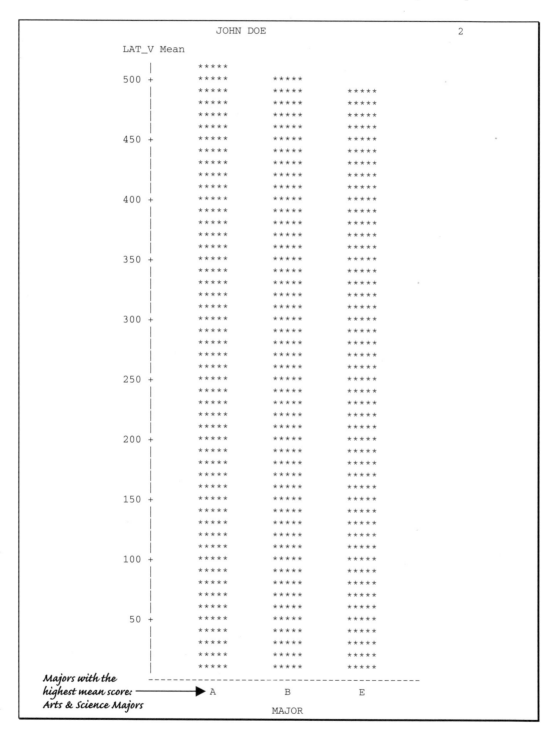

Solution for Chapter 7: Measures of Central Tendency and Variability

7 SOLUTION

Solution to Exercise 7.1: Using PROC UNIVARIATE to Identify Normal, Skewed, and Bimodal Distributions

The SAS Program

```
OPTIONS  LS=80  PS=60;
DATA D1;
   INPUT    SUB_NUM
            LAT_V
            LAT_M
            LAT_A
            MAJOR  $ ;
DATALINES;
01   540   540   540   A
02   510   560   550   A
03   500   520   530   A
04   490   550   530   A
05   520   510   510   A
06   520   500   530   A
07   510   470   520   E
08   530   490   520   B
09   500   460   480   B
10   490   480   500   E
11   510   470   470   B
12   500   450   490   E
13   500   460   540   E
```

Continued on the next page

Continued from the previous page

```
14   480   440   490   B
15   490   430   460   B
16   480   450   480   E
17   470   440   470   E
18   460   450   480   B
;

PROC UNIVARIATE  DATA=D1  PLOT  NORMAL;
   VAR  LAT_V  LAT_M  LAT_A;
   TITLE1  'JOHN DOE';
RUN;
```

The SAS Output

```
                         JOHN DOE                                    1
                 The UNIVARIATE Procedure
                    Variable:  LAT_V
                         Moments

N                          18    Sum Weights                 18
Mean                      500    Sum Observations          9000
Std Deviation      20.5798302    Variance            423.529412
Skewness                    0    Kurtosis            -0.1357639
Uncorrected SS        4507200    Corrected SS              7200
Coeff Variation    4.11596604    Std Error Mean       4.8507125

               Basic Statistical Measures
    Location                    Variability

    Mean     500.0000     Std Deviation           20.57983
    Median   500.0000     Variance               423.52941
    Mode     500.0000     Range                   80.00000
                          Interquartile Range     20.00000

               Tests for Location: Mu0=0

     Test           -Statistic-    -----p Value------

     Student's t    t  103.0776    Pr > |t|     <.0001
     Sign           M         9    Pr >= |M|    <.0001
     Signed Rank    S      85.5    Pr >= |S|    <.0001
```

Continued on the next page

Continued from the previous page

```
                    Tests for Normality

 Test                 --Statistic---    -----p Value------

Shapiro-Wilk          W     0.983895    Pr < W      0.9812
Kolmogorov-Smirnov    D     0.111111    Pr > D     >0.1500
Cramer-von Mises      W-Sq  0.036122    Pr > W-Sq  >0.2500
Anderson-Darling      A-Sq  0.196144    Pr > A-Sq  >0.2500
                   Quantiles (Definition 5)

                    Quantile      Estimate
                    100% Max           540
                    99%                540
                    95%                540
                    90%                530
                    75% Q3             510
                    50% Median         500
                    25% Q1             490
                    10%                470
                     5%                460
                     1%                460
                     0% Min            460
```

```
                         JOHN DOE                                2

                The UNIVARIATE Procedure
                   Variable:  LAT_V

                 Extreme Observations

         ----Lowest----         ----Highest---

         Value      Obs         Value      Obs

           460       18           510       11
           470       17           520        5
           480       16           520        6
           480       14           530        8
           490       15           540        1

     Stem Leaf                     #             Boxplot
       54 0                        1                |
       53 0                        1                |
       52 00                       2                |
       51 000                      3             +-----+
       50 0000                     4             *--+--*
       49 000                      3             +-----+
       48 00                       2                |
       47 0                        1                |
       46 0                        1                |
          ----+----+----+----+
       Multiply Stem.Leaf by 10**+1


                    Normal Probability Plot
    545+                                                 *  +++++
       |                                            *  +++++
       |                                       *  +*+++
       |                                   **+*++
    505+                             **+**++
       |                        *  **+++
       |                    *  +*+++
       |                 +*+++
    465+            +*+++
        +----+----+----+----+----+----+----+----+----+----+
            -2        -1         0        +1        +2
```

```
                              JOHN DOE                                3

                      The UNIVARIATE Procedure
                         Variable:  LAT_M

                              Moments

N                          18    Sum Weights                 18
Mean               481.666667    Sum Observations          8670
Std Deviation      40.0367478    Variance            1602.94118
Skewness           0.70394045    Kurtosis            -0.6566599
Uncorrected SS        4203300    Corrected SS             27250
Coeff Variation    8.31212758    Std Error Mean      9.43675196


                    Basic Statistical Measures

         Location                    Variability

     Mean     481.6667     Std Deviation            40.03675
     Median   470.0000     Variance                     1603
     Mode     450.0000     Range                   130.00000
                           Interquartile Range      60.00000


                    Tests for Location: Mu0=0

        Test           -Statistic-    -----p Value------

        Student's t    t  51.04157    Pr > |t|    <.0001
        Sign           M         9    Pr >= |M|   <.0001
        Signed Rank    S      85.5    Pr >= |S|   <.0001


                       Tests for Normality

     Test                  --Statistic---    -----p Value------

     Shapiro-Wilk          W      0.91626    Pr < W      0.1110
     Kolmogorov-Smirnov    D     0.170182    Pr > D     >0.1500
     Cramer-von Mises      W-Sq  0.091495    Pr > W-Sq   0.1383
     Anderson-Darling      A-Sq  0.564225    Pr > A-Sq   0.1281
```

Continued on the next page

Continued from the previous page

```
                    Quantiles (Definition 5)

                    Quantile      Estimate

                     100% Max          560
                     99%               560
                     95%               560
                     90%               550
                     75% Q3            510
                     50% Median        470
                     25% Q1            450
                     10%               440
                      5%               430
                      1%               430
                      0% Min           430
```

```
                         JOHN DOE                         4

                The UNIVARIATE Procedure
                     Variable:  LAT_M

                   Extreme Observations

          ----Lowest----        ----Highest---

          Value      Obs        Value      Obs

            430       15          510        5
            440       17          520        3
            440       14          540        1
            450       18          550        4
            450       16          560        2


     Stem Leaf                      #             Boxplot
       56 0                         1                |
       54 00                        2                |
       52 0                         1                |
       50 00                        2             +-----+
       48 00                        2             |  +  |
       46 0000                      4             *-----*
       44 00000                     5             +-----+
       42 0                         1                |
          ----+----+----+----+
      Multiply Stem.Leaf by 10**+1


                 Normal Probability Plot
    570+                                                 * +++++
       |                                          *  * +++++
       |                                       * +++++
       |                                    *+*++
       |                               ++*+*
       |                          +*+**
       |                     *  *++*++*
    430+                *  +++++
       +----+----+----+----+----+----+----+----+----+----+
           -2        -1         0        +1        +2
```

```
                          JOHN DOE                             5

                  The UNIVARIATE Procedure
                     Variable:  LAT_A

                          Moments

N                          18    Sum Weights                 18
Mean                      505    Sum Observations          9090
Std Deviation      28.3362167    Variance            802.941176
Skewness                    0    Kurtosis             -1.434561
Uncorrected SS        4604100    Corrected SS             13650
Coeff Variation    5.61113202    Std Error Mean      6.67891033


                 Basic Statistical Measures

        Location                    Variability

    Mean     505.0000     Std Deviation            28.33622
    Median   505.0000     Variance                802.94118
    Mode     480.0000     Range                    90.00000
                          Interquartile Range      50.00000

NOTE: The mode displayed is the smallest of 2 modes with a count
of 3.


                  Tests for Location: Mu0=0

        Test           -Statistic-    -----p Value------

       Student's t    t  75.61114    Pr > |t|     <.0001
       Sign           M         9    Pr >= |M|    <.0001
       Signed Rank    S      85.5    Pr >= |S|    <.0001
```

Continued on the next page

Continued from the previous page

Tests for Normality

Test	--Statistic---		-----p Value------	
Shapiro-Wilk	W	0.932562	Pr < W	0.2154
Kolmogorov-Smirnov	D	0.146166	Pr > D	>0.1500
Cramer-von Mises	W-Sq	0.083658	Pr > W-Sq	0.1795
Anderson-Darling	A-Sq	0.49613	Pr > A-Sq	0.1948

Quantiles (Definition 5)

Quantile	Estimate
100% Max	550
99%	550
95%	550
90%	540
75% Q3	530
50% Median	505
25% Q1	480
10%	470
5%	460
1%	460

```
                          JOHN DOE                             6

                 The UNIVARIATE Procedure
                     Variable:  LAT_A

                 Quantiles (Definition 5)

                  Quantile       Estimate

                   0% Min             460


                   Extreme Observations

             ----Lowest----       ----Highest---

             Value       Obs      Value      Obs

               460        15        530        4
               470        17        530        6
               470        11        540        1
               480        18        540       13
               480        16        550        2


        Stem Leaf                   #              Boxplot
          55 0                      1                 |
          54 00                     2                 |
          53 000                    3              +-----+
          52 00                     2              |     |
          51 0                      1              |     |
          50 0                      1              *--+--*
          49 00                     2              |     |
          48 000                    3              +-----+
          47 00                     2                 |
          46 0                      1                 |
             ----+----+----+----+
         Multiply Stem.Leaf by 10**+1
```

Continued on the next page

Continued from the previous page

```
                  Normal Probability Plot
555+                                         ++*+
   |                                    * *++
   |                              * **++++
525+                           * * +++
   |                          *++++
   |                        *++
495+                    +*+*
   |                 **+*
   |            *+*++
465+       *  +++
    +----+----+----+----+----+----+----+----+----+----+
        -2        -1         0        +1        +2
```

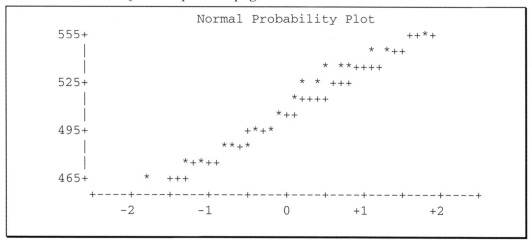

Exercise 7.1, Item 1

Table 7.S1.1

Statistics for the LAT Study

Variable	M	Mdn	Mode	SD	N
LAT Verbal	500.00	500	500	20.58	18
LAT Math	481.67	470	450	40.04	18
LAT Analytical	505.00	505	480	28.34	18

Item 2.A The distribution for LAT Verbal is approximately normal.
I believe that LAT Verbal is approximately normal because it more or less follows the bell-shaped, symmetrical pattern of the normal curve.

Item 2.B The distribution for LAT Math is positively skewed.
I believe that LAT Verbal is positively skewed because one tail is longer than the other, and the longer tail points in the direction of higher values.

Item 2.C The distribution for LAT Analytical is bimodal.
I believe that LAT Verbal is bimodal because it has two "peaks" or "humps."

Solution for Chapter 8: Creating and Modifying Variables and Data Sets

8
SOLUTION

Solution to Exercise 8.1: Working with an Academic Development Questionnaire

The SAS Program

```
OPTIONS  LS=80  PS=60;
DATA D1;
   INPUT  SUB_NUM
          Q1
          Q2
          Q3
          Q4
          SEX  $
          AGE
          CLASS;
DATALINES;
01  6 . 2 7 F 20 1
02  3 2 7 2 M 26 1
03  7 7 1 7 M 19 1
04  5 6 . 5 F 23 2
05  6 7 1 6 M 21 2
06  3 2 6 3 F 25 2
07  5 6 2 5 F 25 3
08  5 6 1 5 F 23 3
09  7 7 1 6 M 31 3
```

Continued on the next page

Continued from the previous page

```
10  5 4 1 4 M 25 4
11  4 5 3 5 F 42 4
12  7 6 1 6 F 38 4
;
PROC PRINT  DATA=D1;
   TITLE1  'JOHN DOE';
RUN;
DATA D2;
   SET D1;

   Q3 = 8 - Q3;

   DEVEL = (Q1 + Q2 + Q3 + Q4) / 4;

   AGE2 = .;
   IF AGE LT 30 THEN AGE2 = 0;
   ELSE IF AGE GE 30 THEN AGE2 = 1;

   CLASS2 = '   .';
   IF CLASS = 1 THEN CLASS2 = 'FRE';
   IF CLASS = 2 THEN CLASS2 = 'SOP';
   IF CLASS = 3 THEN CLASS2 = 'JUN';
   IF CLASS = 4 THEN CLASS2 = 'SEN';

PROC PRINT  DATA=D2;
RUN;
```

The SAS Log

```
NOTE: SAS initialization used:
      real time           19.50 seconds

1     OPTIONS  LS=80  PS=60;
2     DATA D1;
3        INPUT  SUB_NUM
4               Q1
5               Q2
6               Q3
7               Q4
8               SEX  $
9               AGE
10              CLASS;
11    DATALINES;

NOTE: The data set WORK.D1 has 12 observations and 8 variables.
NOTE: DATA statement used:
      real time           1.52 seconds


24    ;
25    PROC PRINT  DATA=D1;
26       TITLE1  'JOHN DOE';
27    RUN;

NOTE: There were 12 observations read from the data set WORK.D1.
NOTE: PROCEDURE PRINT used:
      real time           1.04 seconds
```

Continued on the next page

Continued from the previous page

```
28
29   DATA D2;
30      SET D1;31
32      Q3 = 8 - Q3;
33
34      DEVEL = (Q1 + Q2 + Q3 + Q4) / 4;
35
36      AGE2 = .;
37      IF AGE LT 30 THEN AGE2 = 0;
38      ELSE IF AGE GE 30 THEN AGE2 = 1;
39
40      CLASS2 = '  .';
41      IF CLASS = 1 THEN CLASS2 = 'FRE';
42      IF CLASS = 2 THEN CLASS2 = 'SOP';
43      IF CLASS = 3 THEN CLASS2 = 'JUN';
44      IF CLASS = 4 THEN CLASS2 = 'SEN';
45

NOTE: Missing values were generated as a result of performing an
operation on
      missing values.
      Each place is given by: (Number of times) at
(Line):(Column).
      1 at 32:11   1 at 34:16   2 at 34:21   2 at 34:26   2 at
34:32
NOTE: There were 12 observations read from the data set WORK.D1.
NOTE: The data set WORK.D2 has 12 observations and 11 variables.
NOTE: DATA statement used:
      real time           0.53 seconds


46   PROC PRINT  DATA=D2;
47   RUN;

NOTE: There were 12 observations read from the data set WORK.D2.
NOTE: PROCEDURE PRINT used:
      real time           0.16 seconds
```

The SAS Output

JOHN DOE 1

Obs	SUB_NUM	Q1	Q2	Q3	Q4	SEX	AGE	CLASS
1	1	6	.	2	7	F	20	1
2	2	3	2	7	2	M	26	1
3	3	7	7	1	7	M	19	1
4	4	5	6	.	5	F	23	2
5	5	6	7	1	6	M	21	2
6	6	3	2	6	3	F	25	2
7	7	5	6	2	5	F	25	3
8	8	5	6	1	5	F	23	3
9	9	7	7	1	6	M	31	3
10	10	5	4	1	4	M	25	4
11	11	4	5	3	5	F	42	4
12	12	7	6	1	6	F	38	4

JOHN DOE 2

Obs	SUB_NUM	Q1	Q2	Q3	Q4	SEX	AGE	CLASS	DEVEL	AGE2	CLASS2
1	1	6	.	6	7	F	20	1	.	0	FRE
2	2	3	2	1	2	M	26	1	2.00	0	FRE
3	3	7	7	7	7	M	19	1	7.00	0	FRE
4	4	5	6	.	5	F	23	2	.	0	SOP
5	5	6	7	7	6	M	21	2	6.50	0	SOP
6	6	3	2	2	3	F	25	2	2.50	0	SOP
7	7	5	6	6	5	F	25	3	5.50	0	JUN
8	8	5	6	7	5	F	23	3	5.75	0	JUN
9	9	7	7	7	6	M	31	3	6.75	1	JUN
10	10	5	4	7	4	M	25	4	5.00	0	SEN
11	11	4	5	5	5	F	42	4	4.75	1	SEN
12	12	7	6	7	6	F	38	4	6.50	1	SEN

9 SOLUTION

Solution for Chapter 9: *z* Scores

Solution to Exercise 9.1: Satisfaction with Academic Development and the Social Environment Among College Students

The SAS Program

```
OPTIONS  LS=80  PS=60;
DATA D1;
   INPUT  SUB_NUM
          NAME  $
          ACADEM
          SOCIAL;
DATALINES;
01  Fred       25   34
02  Susan      16   30
03  Marsha      4   42
04  Charles    12   24
05  Paul       28   39
06  Cindy      15   29
07  Jack       21   27
08  Cathy       8   36
09  George     23   21
10  John        6   32
11  Marie      10   19
12  Emmett     19   33
;
```

Continued on the next page

Continued from the previous page

```
PROC MEANS DATA=D1  VARDEF=N  N  MEAN  STD  MIN  MAX;
   VAR  ACADEM  SOCIAL;
   TITLE1  'JOHN DOE';
RUN;
DATA D2;
   SET D1;
   ACADEM_Z = (ACADEM - 15.58) / 7.45;
   SOCIAL_Z = (SOCIAL - 30.50) / 6.68;

PROC PRINT DATA=D2;
   VAR  NAME  ACADEM  SOCIAL  ACADEM_Z  SOCIAL_Z;
   TITLE 'JOHN DOE';
RUN;

PROC MEANS DATA=D2  VARDEF=N  N  MEAN  STD  MIN  MAX;
   VAR  ACADEM_Z  SOCIAL_Z;
   TITLE1  'JOHN DOE';
RUN;
```

The SAS Output

```
                          JOHN DOE

                     The MEANS Procedure

Variable    N        Mean       Std Dev       Minimum       Maximum
--------------------------------------------------------------------
ACADEM     12   15.5833333    7.4549573     4.0000000    28.0000000
SOCIAL     12   30.5000000    6.6770752    19.0000000    42.0000000
--------------------------------------------------------------------
```

```
                              JOHN DOE

  Obs    NAME        ACADEM    SOCIAL     ACADEM_Z     SOCIAL_Z

    1    Fred          25        34        1.26443      0.52395
    2    Susan         16        30        0.05638     -0.07485
    3    Marsha         4        42       -1.55436      1.72156
    4    Charles       12        24       -0.48054     -0.97305
    5    Paul          28        39        1.66711      1.27246
    6    Cindy         15        29       -0.07785     -0.22455
    7    Jack          21        27        0.72752     -0.52395
    8    Cathy          8        36       -1.01745      0.82335
    9    George        23        21        0.99597     -1.42216
   10    John           6        32       -1.28591      0.22455
   11    Marie         10        19       -0.74899     -1.72156
   12    Emmett        19        33        0.45906      0.37425

                             JOHN DOE

                        The MEANS Procedure

Variable    N          Mean       Std Dev       Minimum       Maximum
----------------------------------------------------------------------
ACADEM_Z   12    0.000447427    1.0006654    -1.5543624     1.6671141
SOCIAL_Z   12   -1.85037E-17    0.9995622    -1.7215569     1.7215569
----------------------------------------------------------------------
```

Exercise 9.1: Answers to Questions

1. **Question:** The third SAS procedure in your program should have performed PROC MEANS on ACADEM_Z and SOCIAL_Z. Based on the means and standard deviations for these *z*-score variables, is there reason to believe that the *z*-score variables were created correctly? Explain your answer.

   Answer: Yes, the two *z*-score variables were created correctly. When a *z*-score variable is created correctly, we expect its mean to be approximately zero. The output from PROC MEANS shows that the mean for ACADEM_Z 0.000447427, which was close to zero. The mean for SOCIAL_Z –1.85037E–17, which, in scientific notation, indicates that the mean –0.0000000000000000185037, which again was close to zero.

When a *z*-score variable is created correctly, we also expect the sample standard deviation to be approximately 1. The output shows that the standard deviation for ACADEM_Z 1.0006654, and the standard deviation for SOCIAL_Z 0.9995622. Both of these standard deviations were close to 1.

2. **Question:** Marsha's raw score on ACADEM was 4 (Marsha was observation 3). What was the relative standing of this score within the sample? Explain your answer. (Hint: Your answer should refer to the *z* score that corresponds to this raw score as stated earlier in Chapter 9 of the *Student Guide*).

   Answer. Marsha's score was 1.55 standard deviations below the mean. I know that her score was below the mean because her *z* score was negative in sign. I know that her score was 1.55 standard deviations from the mean because the absolute value of the *z* score was 1.55.

3. **Question:** Jack's raw score on ACADEM was 21 (Jack was observation 7). What was the relative standing of this score within the sample? Explain your answer.

   Answer. Jack's score was 0.73 standard deviations above the mean. I know that his score was above the mean because his *z* score was positive in sign. I know that his score was 0.73 standard deviations from the mean because the absolute value of the *z* score was 0.73.

4. **Question:** Compared to the other subjects, did Fred (observation 1) score higher on the academic development scale or on the social environment scale? Explain your answer. (Hint: Your answer should refer to the *z* scores for these variables as stated earlier in Chapter 9 of the *Student Guide*).

   Answer. Compared to the other students, Fred scored higher on the academic development scale than on the social environment scale. I know this because both *z* scores were positive, and his *z* score on the academic development scale (1.26) was higher than his score on the social development scale (0.52).

5. **Question:** Compared to the other subjects, did Marie (observation 11) score higher on the academic development scale or on the social environment scale? Explain your answer.

   Answer. Compared to the other subjects, Marie scored higher on the academic development scale than on the social environment scale. I know this because both *z* scores were negative, and her *z* score on the academic development scale (–0.75) was closer to zero than her score on the social environment scale (–1.72).

6. **Question:** Compared to the other subjects, did Susan (observation 2) score higher on the academic development scale or on the social environment scale? Explain your answer.

   Answer. Compared to the other subjects, Susan scored higher on the academic development scale than on the social environment scale. I know this because her *z* score on the academic development scale was a positive value (0.06), while her *z* score on the social environment scale was a negative value (–0.07).

Solution for Chapter 10: Bivariate Correlation

10 SOLUTION

Solution to Exercise 10.1: Correlational Study of Drinking and Driving Behavior

The SAS Program

```
OPTIONS  LS=80  PS=60 ;
DATA D1;
   INPUT  SUB_NUM
          MC_T1
          PC_T1
          DD_T1
          DD_T2;
DATALINES;
01   24   8 1 3
02   12  12 3 4
03    4   4 4 5
04   16  12 2 2
05   28   . 1 1
06    8  16 3 3
07   12   8 3 5
08    8   8 5 6
09   16  16 4 4
10   16  16 4 5
11   12  12 4 6
12   20  16 5 2
13   28  24 5 3
```

Continued on the next page

Continued from the previous page

```
14   20 20 5 5
15   20 24 6 4
16   16 20 6 6
17   16 16 6 7
18    4 12 6 8
19   24 28 7 5
20   20 24 7 6
21    4 20 8 8
;

PROC PLOT  DATA=D1;
   PLOT  DD_T2*MC_T1;
   TITLE1 'JOHN DOE';
RUN;

PROC CORR  DATA=D1;
   VAR  MC_T1  PC_T1  DD_T1  DD_T2;
   TITLE1 'JOHN DOE';
RUN;
```

The SAS Output

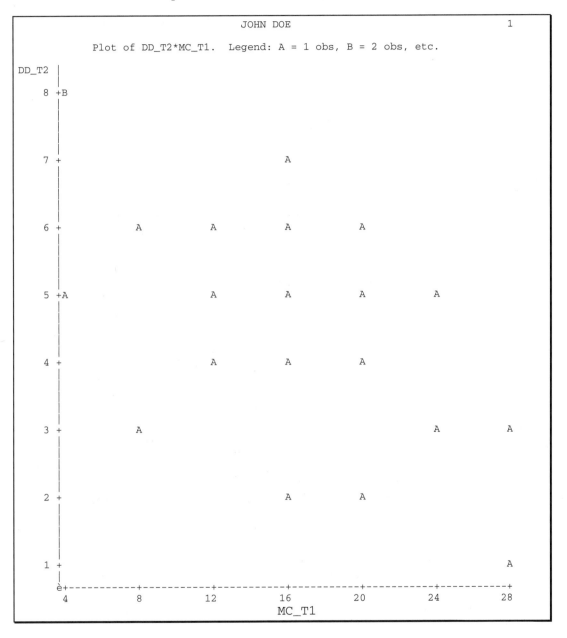

```
                                    JOHN DOE                                   1

            Plot of DD_T2*MC_T1.  Legend: A = 1 obs, B = 2 obs, etc.

DD_T2 |
      |
    8 +B
      |
      |
      |
      |
      |
    7 +                                    A
      |
      |
      |
      |
      |
    6 +            A           A           A           A
      |
      |
      |
      |
      |
    5 +A                       A           A           A           A
      |
      |
      |
      |
      |
    4 +                        A           A           A
      |
      |
      |
      |
      |
    3 +            A                                               A           A
      |
      |
      |
      |
      |
    2 +                                    A           A
      |
      |
      |
      |
      |
    1 +                                                                        A
      |
      è+-----------+-----------+-----------+-----------+-----------+-----------+
       4           8          12          16          20          24          28
                                           MC_T1
```

```
                         JOHN DOE                                 2


                   The CORR Procedure

     4  Variables:     MC_T1     PC_T1     DD_T1     DD_T2


                   Simple Statistics

Variable N         Mean    Std Dev         Sum    Minimum    Maximum

MC_T1     21   15.61905    7.36530   328.00000    4.00000   28.00000
PC_T1     20   15.80000    6.42036   316.00000    4.00000   28.00000
DD_T1     21    4.52381    1.91361    95.00000    1.00000    8.00000
DD_T2     21    4.66667    1.90613    98.00000    1.00000    8.00000


              Pearson Correlation Coefficients
                 Prob > |r| under H0: Rho=0
                   Number of Observations

                 MC_T1          PC_T1          DD_T1          DD_T2

    MC_T1      1.00000        0.55954       -0.19799       -0.59358
                               0.0103         0.3896         0.0046
                    21             20             21             21

    PC_T1      0.55954        1.00000        0.65765        0.03456
                0.0103                        0.0016         0.8850
                    20             20             20             20

    DD_T1     -0.19799        0.65765        1.00000        0.70823
                0.3896         0.0016                        0.0003
                    21             20             21             21

    DD_T2     -0.59358        0.03456        0.70823        1.00000
                0.0046         0.8850         0.0003
                    21             20             21             21
```

Report Summarizing the Results of the Analysis

A) <u>Statement of the research question</u>: The purpose of this study was to determine whether moral commitment to the legal norm is correlated with drinking and driving at Time 2.

B) <u>Statement of the research hypothesis</u>: There will be a negative relationship between moral commitment to the legal norm and drinking and driving behavior.

C) <u>Nature of the variables</u>: This analysis involved one predictor variable and one criterion variable.

- The predictor variable was moral commitment to the legal norm. This was a multi-value variable and was assessed on an interval scale.
- The criterion variable was the number times the subject drove a car after drinking alcoholic beverages during a 1-month period. This was a multi-value variable and was assessed on a ratio scale.

D) <u>Statistical test</u>: Pearson product-moment correlation coefficient.

E) <u>Statistical null hypothesis (H_0)</u>: $\rho = 0$; In the population, the correlation between moral commitment to the legal norm and drinking and driving behavior is equal to zero.

F) <u>Statistical alternative hypothesis (H_1)</u>: $\rho \neq 0$; In the population, the correlation between moral commitment to the legal norm and drinking and driving behavior is not equal to zero.

G) <u>Obtained statistic</u>: $r = -.59$

H) <u>Obtained probability (p) value</u>: $p = .0046$

I) <u>Conclusion regarding the statistical null hypothesis</u>: Reject the null hypothesis.

J) Conclusion regarding the research hypothesis: These findings provide support for the study's research hypothesis.

K) Coefficient of determination: .35.

L) Formal description of the results for a paper: Results were analyzed by computing a Pearson product-moment correlation coefficient. This analysis revealed a significant negative correlation between moral commitment to the legal norm and drinking and driving behavior, r = -.59, p = .0046. The nature of the correlation coefficient showed that subjects who scored higher on moral commitment tended to drink and drive less frequently than those who scored lower on moral commitment. The coefficient of determination showed that moral commitment accounted for 35 percent of the variance in drinking and driving behavior.

Solution for Chapter 11: Bivariate Regression

11 SOLUTION

Solution to Exercise 11.1: Predicting Current Drinking and Driving Behavior from Previous Behavior

The SAS Program

```
OPTIONS  LS=80  PS=60 ;
DATA D1;
   INPUT  SUB_NUM
          MC_T1
          PC_T1
          DD_T1
          DD_T2;
DATALINES;
01  24  8 1 3
02  12 12 3 4
03   4  4 4 5
04  16 12 2 2
05  28  . 1 1
06   8 16 3 3
07  12  8 3 5
08   8  8 5 6
09  16 16 4 4
10  16 16 4 5
11  12 12 4 6
12  20 16 5 2
```

Continued on the next page

Continued from the previous page

```
13   28 24 5 3
14   20 20 5 5
15   20 24 6 4
16   16 20 6 6
17   16 16 6 7
18    4 12 6 8
19   24 28 7 5
20   20 24 7 6
21    4 20 8 8
;

PROC PLOT  DATA=D1;
   PLOT  DD_T2*DD_T1;
   TITLE1 'JOHN DOE';
RUN;

PROC REG  DATA=D1;
   MODEL  DD_T2=DD_T1  /  STB  P;
   TITLE1 'JOHN DOE';
RUN;
```

The SAS Output

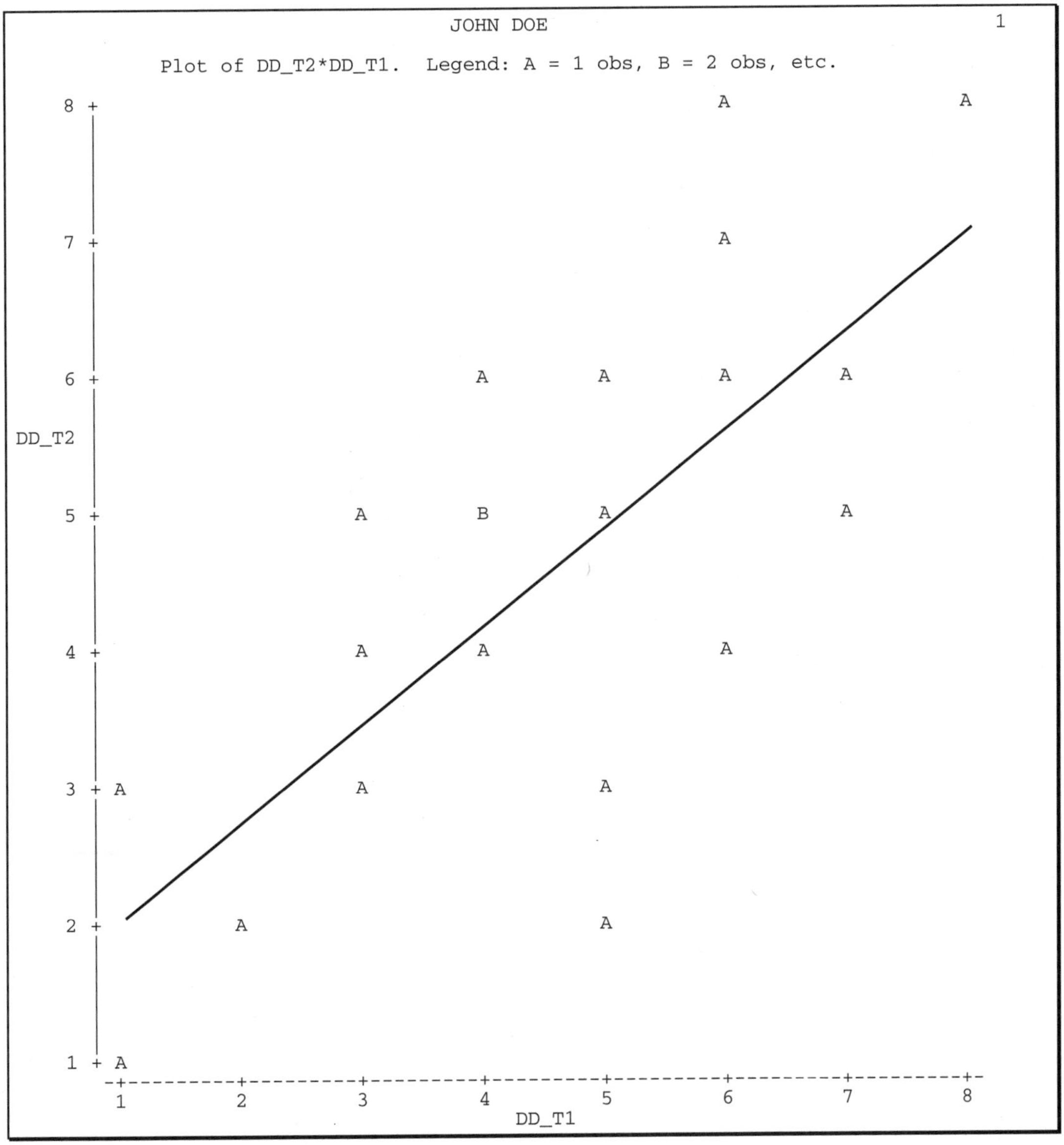

```
                              JOHN DOE                                  2

                          The REG Procedure
                           Model: MODEL1
                     Dependent Variable: DD_T2

                        Analysis of Variance

                                   Sum of           Mean
Source                  DF        Squares         Square    F Value    Pr > F

Model                    1       36.44885       36.44885      19.12    0.0003
Error                   19       36.21782        1.90620
Corrected Total         20       72.66667


         Root MSE               1.38065    R-Square      0.5016
         Dependent Mean         4.66667    Adj R-Sq      0.4754
         Coeff Var             29.58541


                         Parameter Estimates

                   Parameter    Standard                            Standardized
Variable     DF     Estimate       Error    t Value    Pr > |t|         Estimate

Intercept     1      1.47529     0.78957       1.87      0.0772                0
DD_T1         1      0.70546     0.16133       4.37      0.0003          0.70823
```

JOHN DOE 3

The REG Procedure
Model: MODEL1
Dependent Variable: DD_T2

Output Statistics

Obs	Dep Var DD_T2	Predicted Value	Residual
1	3.0000	2.1808	0.8192
2	4.0000	3.5917	0.4083
3	5.0000	4.2971	0.7029
4	2.0000	2.8862	-0.8862
5	1.0000	2.1808	-1.1808
6	3.0000	3.5917	-0.5917
7	5.0000	3.5917	1.4083
8	6.0000	5.0026	0.9974
9	4.0000	4.2971	-0.2971
10	5.0000	4.2971	0.7029
11	6.0000	4.2971	1.7029
12	2.0000	5.0026	-3.0026
13	3.0000	5.0026	-2.0026
14	5.0000	5.0026	-0.002601
15	4.0000	5.7081	-1.7081
16	6.0000	5.7081	0.2919
17	7.0000	5.7081	1.2919
18	8.0000	5.7081	2.2919
19	5.0000	6.4135	-1.4135
20	6.0000	6.4135	-0.4135
21	8.0000	7.1190	0.8810

Sum of Residuals	0
Sum of Squared Residuals	36.21782
Predicted Residual SS (PRESS)	43.13137

Report Summarizing the Results of the Analysis

A) Statement of the research question: The purpose of this study was to determine whether the regression coefficient representing the relationship between drinking and driving behavior at Time 1 (during the month of June) and drinking and driving behavior at Time 2 (during the month of July) is significantly different from zero.

B) Statement of the research hypothesis: There will be a positive relationship between drinking and driving behavior at Time 1 and drinking and driving behavior at Time 2.

C) Nature of the variables: This analysis involved one predictor variable and one criterion variable.

- The predictor variable was drinking and driving behavior at Time 1. This was a multi-value variable and was assessed on a ratio scale.
- The criterion variable was drinking and driving behavior at Time 2. This was a multi-value variable and was also assessed on a ratio scale.

D) Statistical procedure: Linear bivariate regression.

E) Statistical null hypothesis (H_0): $b = 0$; In the population, the regression coefficient representing the relationship between drinking and driving behavior at Time 1 and drinking and driving behavior at Time 2 is equal to zero.

F) Statistical alternative hypothesis (H_1): $b \neq 0$; In the population, the regression coefficient representing the relationship between drinking and driving behavior at Time 1 and drinking and driving behavior at Time 2 is not equal to zero.

G) Obtained statistic: $b = .705$, $t\ (19) = 4.37$

H) <u>Obtained probability (p) value</u>: $p = .0003$

I) <u>Conclusion regarding the statistical null hypothesis</u>: Reject the null hypothesis.

J) <u>Conclusion regarding the research hypothesis</u>: These findings provide support for the study's research hypothesis.

K) <u>Coefficient of determination</u>: .50

L) <u>Formal description of the results for a paper</u>: Results were analyzed by using linear regression to regress drinking and driving behavior at Time 2 on drinking and driving behavior at Time 1. This analysis revealed a significant regression coefficient, $b = .705$, $t(19) = 4.37$, $p = .0003$. The nature of the regression coefficient showed that, on the average, an increase of .705 incidents of drinking and driving behavior at Time 2 was associated with every 1-unit increase in drinking and driving behavior at Time 1. The analysis showed that drinking and driving behavior at Time 1 accounted for 50% of the variance in drinking and driving behavior at Time 2.

Computing a predicted value of Y that is associated with a low value of X

$$Y' = b\,(X) + a$$

$$Y' = .705(X) + 1.475$$

$$Y' = .705(2) + 1.475$$

$$Y' = 1.410 + 1.475$$

$$Y' = 2.885$$

$$Y' = 2.89$$

Computing a predicted value of Y that is associated with a high value of X

$Y' = b\,(X) + a$

$Y' = .705(X) + 1.475$

$Y' = .705(7) + 1.475$

$Y' = 4.935 + 1.475$

$Y' = 6.410$

$Y' = 6.41$

12 SOLUTION

Solution for Chapter 12: Single-Sample *t* Test

Solution to Exercise 12.1: Answering SAT Reading Comprehension Questions Without the Passages

The SAS Program

```
OPTIONS  LS=80  PS=60;
DATA D1;
   INPUT  SUB_NUM
          SCORE;
DATALINES;
01  49
02  46
03  51
04  53
05  55
06  50
07  43
08  53
09  49
10  51
11  49
12  47
13  57
```

Continued on the next page

Continued from the previous page

```
14  51
15  47
16  45
17  45
;
PROC TTEST  DATA=D1  H0=20  ALPHA=0.05;
   VAR SCORE;
   TITLE1  'JANE DOE';
RUN;
```

The SAS Output

```
                              JANE DOE                                    1

                         The TTEST Procedure

                              Statistics

              Lower CL          Upper CL   Lower CL               Upper CL
Variable    N    Mean     Mean     Mean    Std Dev    Std Dev    Std Dev

SCORE      17   47.537  49.471   51.404    2.8005     3.7603     5.7229


                              Statistics

           Variable    Std Err     Minimum     Maximum

           SCORE         0.912          43          57


                               T-Tests

            Variable        DF     t Value     Pr > |t|

            SCORE           16       32.31      <.0001
```

Computing the Index of Effect Size

$$d = \frac{|\overline{X} - \mu_0|}{s_X}$$

$$d = \frac{49.471 - 20}{3.7603}$$

$$d = \frac{29.471}{3.7603}$$

$$d = 7.837$$

$$d = 7.83$$

Report Summarizing the Results of the Analysis

A) <u>Statement of the research question</u>: The purpose of this study was to determine whether subjects answering SAT reading comprehension items without passages would perform at a level that is higher than the level expected from random responses.

B) <u>Statement of the research hypothesis</u>: Subjects answering SAT Reading Comprehension items without passages will perform at a level that is higher than the level expected from random responses.

C) <u>Nature of the variable</u>: The criterion variable was "number of correct responses," which is the number of items that the subject answered correctly out of a possible 100 items. This was a multi-value variable and was assessed on a ratio scale.

D) Statistical test: Single-sample t test.

E) Statistical null hypothesis (H_0): $\mu = 20$; In the population, the average number of correct responses is equal to 20 out of 100 (the number expected from random responses).

F) Statistical alternative hypothesis (H_1): $\mu \neq 20$; In the population, the average number of correct responses is not equal to 20 out of 100.

G) Obtained statistic: $t = 32.31$

H) Obtained probability (p) value: $p < .0001$

I) Conclusion regarding the statistical null hypothesis: Reject the null hypothesis.

J) Confidence interval. The sample mean on the criterion variable (number of correct responses) was 49.471. The 95 percent confidence interval for the mean extended from 47.537 to 51.404.

K) Effect size. $d = 7.83$.

L) Conclusion regarding the research hypothesis: These findings provide support for the study's research hypothesis.

M) Formal description of results for a paper: Results were analyzed using a single-sample t test. This analysis revealed a significant t value, $t(16) = 32.31$, $p < .0001$. In the sample, the mean number of correct responses was 49.471 ($SD = 3.7603$), which was significantly higher than the 20 correct answers that would have been expected from random responding. The 95 percent confidence interval for the mean extended from 47.537 to 51.404. The effect size was computed as $d = 7.83$. According to Cohen's (1969) guidelines, this represents a relatively large effect.

13 SOLUTION

Solution for Chapter 13: Independent-Samples *t* Test

Solution to Exercise 13.1: Sex Differences in Sexual Jealousy

The SAS Program

```
OPTIONS  LS=80  PS=60;
DATA D1;
   INPUT  SUB_NUM
          SEX  $
          DISTRESS;
DATALINES;
01  M  22
02  M  25
03  M  23
04  M  24
05  M  20
06  M  28
07  M  27
08  M  23
09  M  23
10  M  24
11  M  26
12  M  26
13  M  25
14  M  21
15  M  22
```

Continued on the next page

Continued from the previous page

```
16  M  23
17  M  24
18  M  25
19  F  22
20  F  22
21  F  25
22  F  18
23  F  23
24  F  24
25  F  19
26  F  20
27  F  20
28  F  20
29  F  21
30  F  21
31  F  21
32  F  19
33  F  19
34  F  22
35  F  23
36  F  21
;

PROC TTEST  DATA=D1  ALPHA=0.05;
   CLASS  SEX;
   VAR DISTRESS;
   TITLE1  'JANE DOE';
RUN;
```

The SAS Output

JANE DOE 1

The TTEST Procedure

Statistics

Variable	Class	N	Lower CL Mean	Mean	Upper CL Mean	Lower CL Std Dev	Std Dev
DISTRESS	F	18	20.179	21.111	22.044	1.4071	1.8752
DISTRESS	M	18	22.914	23.944	24.975	1.5544	2.0714
DISTRESS	Diff (1-2)		-4.172	-2.833	-1.495	1.5981	1.9758

Statistics

Variable	Class	Upper CL Std Dev	Std Err	Minimum	Maximum
DISTRESS	F	2.8112	0.442	18	25
DISTRESS	M	3.1054	0.4882	20	28
DISTRESS	Diff (1-2)	2.5886	0.6586		

T-Tests

Variable	Method	Variances	DF	t Value	Pr > |t|
DISTRESS	Pooled	Equal	34	-4.30	0.0001
DISTRESS	Satterthwaite	Unequal	33.7	-4.30	0.0001

Equality of Variances

Variable	Method	Num DF	Den DF	F Value	Pr > F
DISTRESS	Folded F	17	17	1.22	0.6862

Computing the Index of Effect Size

$$d = \frac{| \overline{X}_1 - \overline{X}_2 |}{s_p}$$

$$d = \frac{| 21.111 - 23.944 |}{1.9758}$$

$$d = \frac{| -2.833 |}{1.9758}$$

$$d = \frac{2.833}{1.9758}$$

$$d = 1.4338$$

$$d = 1.43$$

Report Summarizing the Results of the Analysis

A) Statement of the research question: The purpose of this study was to ask a group of subjects to imagine how they would feel if they learned that their partners had had intercourse with someone else, and then determine whether male subjects would score higher than female subjects on a measure of psychological distress.

B) Statement of the research hypothesis: When asked to imagine how they would feel if they learned that their partner had had sexual intercourse with someone else, male subjects will display higher mean scores than female subjects on a measure of psychological distress.

C) **Nature of the variables:** This analysis involved one predictor variable and one criterion variable.

- The predictor variable was subject sex. This was a dichotomous variable and was assessed on a nominal scale.
- The criterion variable was subjects' scores on a 4-item measure of distress. This was a multi-value variable and was assessed on an interval scale.

D) **Statistical test:** Independent-samples $\underline{t}$ test.

E) **Statistical null hypothesis** **(H_0):** $\mu_1 = \mu_2$; In the population, there is no difference between male subjects and female subjects with respect to their mean scores for the criterion variable (the measure of distress).

F) **Statistical alternative hypothesis** **(H_1):** $\mu_1 \neq \mu_2$; In the population, there is a difference between male subjects and female subjects with respect to their mean scores for the criterion variable (the measure of distress).

G) **Obtained statistic:** $\underline{t} = -4.30$

H) **Obtained probability (p) value:** $\underline{p} = .0001$

I) **Conclusion regarding the statistical null hypothesis:** Reject the null hypothesis.

J) **Confidence interval:** Subtracting the mean of the male subjects from the mean of the female subjects resulted in an observed difference of -2.833. The 95 percent confidence interval for this difference extended from -4.172 to -1.495.

K) **Effect size:** $\underline{d} = 1.43$.

L) Conclusion regarding the research hypothesis: These findings provide support for the study's research hypothesis.

M) Formal description of results for a paper: Results were analyzed using an independent-samples t test. This analysis revealed a significant difference between the two groups, $t(34) = -4.30$, $p = .0001$. The sample means are displayed in Figure 13.S1.1, which shows that male subjects scored significantly higher on distress compared to female subjects (for males, $M = 23.94$, $SD = 2.07$; for females, $M = 21.11$, $SD = 1.88$). The observed difference between the means was -2.83, and the 95 percent confidence interval for the difference between means extended from -4.17 to -1.50. The effect size was computed as $d = 1.43$. According to Cohen's (1969) guidelines, this represents a relatively large effect.

N) Figure representing the results: See Figure 13.S1.1.

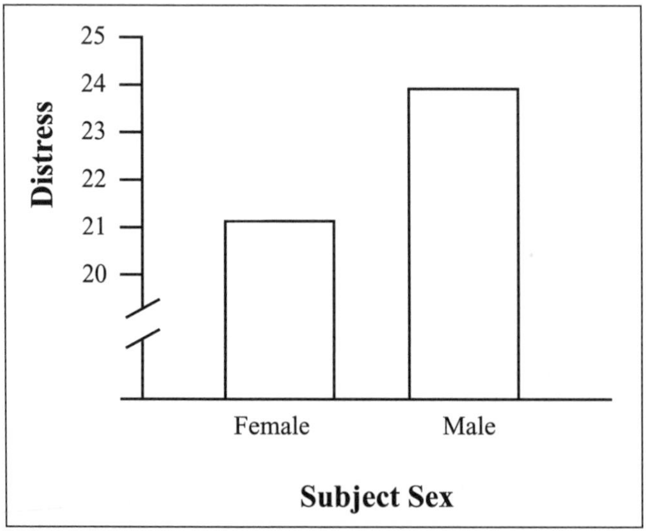

Figure 13.S1.1. Mean scores for the measure of psychological distress as a function of subject sex.

14
SOLUTION

Solution for Chapter 14: Paired-Samples *t* Test

Solution to Exercise 14.1: Perceived Problem Seriousness as a Function of Time of Day

The SAS Program

```
OPTIONS  LS=80  PS=60;
DATA D1  ;
   INPUT  SUB_NUM
          AFTNOON
          MORNING;
DATALINES;
01  16  13
02  14  14
03  12   9
04  16  12
05  13  12
06  11  13
07  17  13
08  10  10
09  14  11
10  15  11
11  14   8
12  13  11
13  15  11
```

Continued on the next page

Continued from the previous page

```
14  13   8
15  13  10
16  11  10
;
PROC MEANS  DATA=D1;
   VAR  AFTNOON  MORNING;
   TITLE1  'JOHN DOE';
RUN;

PROC TTEST  DATA=D1  H0=0  ALPHA=0.05;
   PAIRED  AFTNOON*MORNING;
RUN;
```

The SAS Output

```
                                JOHN DOE                                     1

                          The MEANS Procedure

Variable     N         Mean       Std Dev       Minimum       Maximum
-------------------------------------------------------------------------
AFTNOON     16    13.5625000    1.9653244    10.0000000    17.0000000
MORNING     16    11.0000000    1.7888544     8.0000000    14.0000000
-------------------------------------------------------------------------
```

```
                          JOHN DOE                                   2

                    The TTEST Procedure

                         Statistics

                         Lower CL            Upper CL   Lower CL
Difference          N      Mean      Mean      Mean     Std Dev   Std Dev

AFTNOON - MORNING  16    1.4453    2.5625    3.6797    1.5488    2.0966

                         Statistics

                        Upper CL
   Difference           Std Dev    Std Err    Minimum     Maximum

   AFTNOON - MORNING     3.2449     0.5242         -2           6


                            T-Tests

            Difference              DF    t Value    Pr > |t|
            AFTNOON - MORNING       15       4.89      0.0002
```

Computing the Index of Effect Size

$$d = \frac{| \overline{X}_1 - \overline{X}_2 |}{s_D}$$

$$d = \frac{| 13.5625 - 11.000 |}{2.0966}$$

$$d = \frac{| 2.5625 |}{2.0966}$$

$$d = \frac{2.5625}{2.0966}$$

$$d = 1.2222$$

$$d = 1.22$$

Report Summarizing the Results of the Analysis

A) Statement of the research question: The purpose of this study was to determine whether there is a difference between mean perceived problem seriousness scores obtained during the afternoon and those obtained in the morning.

B) Statement of the research hypothesis: Mean perceived problem seriousness scores obtained during the afternoon will be higher (rated as more serious) than those obtained in the morning.

C) Nature of the variables: This analysis involved one predictor variable and one criterion variable.

- The predictor variable was time of day. This was a dichotomous variable, was assessed on a nominal scale, and included two conditions: afternoon and morning.

- The criterion variable was subjects' scores on a 3-item measure of perceived problem seriousness. This was a multi-value variable and was assessed on an interval scale.

D) Statistical test: Paired-samples t test.

E) Statistical null hypothesis (H_0): $\mu_1 = \mu_2$; In the population, there is no difference between the afternoon condition and the morning condition with respect to mean scores for the criterion variable (perceived problem seriousness).

F) Statistical alternative hypothesis (H_1): $\mu_1 \neq \mu_2$; In the population, there is a difference between the afternoon condition and the morning condition with respect to mean scores for the criterion variable (perceived problem seriousness).

G) Obtained statistic: $t = 4.89$

H) Obtained probability (p) value: $p = .0002$

I) Conclusion regarding the statistical null hypothesis: Reject the null hypothesis.

J) Confidence interval: Subtracting the mean of the morning condition from the mean of the afternoon condition resulted in an observed difference of 2.56. The 95 percent confidence interval for this difference extended from 1.45 to 3.68.

K) Effect size: $d = 1.22$.

L) Conclusion regarding the research hypothesis: These findings provide support for the study's research hypothesis.

M) Formal description of results for a paper: Results were analyzed using a paired-samples t test. This analysis revealed a statistically significant difference between the two conditions, $t(15) = 4.89$, $p = .0002$. The sample means are displayed in Figure 14.S1.1, which shows that the mean perceived problem seriousness score obtained under the afternoon condition was significantly higher than the mean score obtained under the morning condition (for afternoon, $M = 13.56$, $SD = 1.97$; for morning, $M = 11.00$, $SD = 1.79$). The observed difference between the means was 2.56, and the 95 percent confidence interval for the difference between means extended from 1.45 to 3.68. The effect size was computed as $d = 1.22$. According to Cohen's (1969) guidelines, this represents a relatively large effect.

N) <u>Figure representing the results</u>:

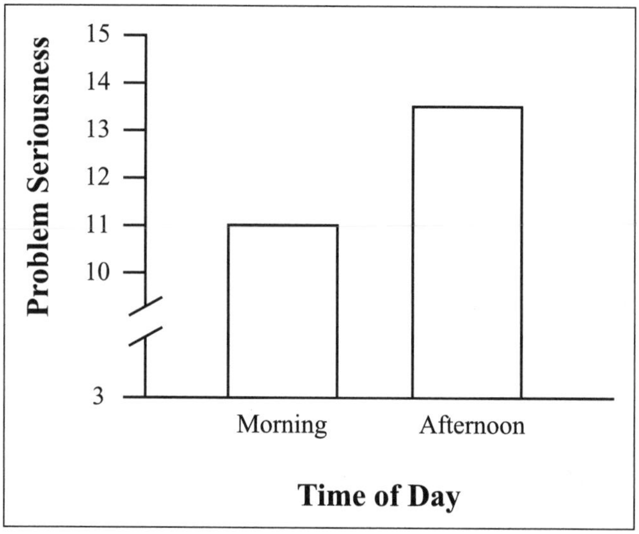

Figure 14.S1.1. Mean scores for perceived problem seriousness as a function of time of day.

Solution for Chapter 15: One-Way ANOVA with One Between-Subjects Factor

15 SOLUTION

Solution to Exercise 15.1: The Effect of Misleading Suggestions on the Creation of False Memories

The SAS Program

```
OPTIONS  LS=80  PS=60;
DATA D1;
   INPUT  SUB_NUM
          NUM_EXP  $
          CONFID      ;
DATALINES;
01  0_EXP  1.00
02  0_EXP  1.25
03  0_EXP  2.00
04  0_EXP  1.75
05  0_EXP  2.75
06  0_EXP  3.25
07  0_EXP  4.50
08  2_EXP  2.75
09  2_EXP  3.00
10  2_EXP  4.00
11  2_EXP  5.50
12  2_EXP  5.50
13  2_EXP  6.25
```

Continued on the next page

Continued from the previous page

```
14  2_EXP  6.75
15  4_EXP  4.75
16  4_EXP  3.00
17  4_EXP  4.75
18  4_EXP  5.75
19  4_EXP  7.00
20  4_EXP  6.00
21  4_EXP  3.00
;

PROC GLM  DATA=D1;
   CLASS  NUM_EXP;
   MODEL  CONFID = NUM_EXP;
   MEANS  NUM_EXP;
   MEANS  NUM_EXP / TUKEY  CLDIFF  ALPHA=0.05;
   TITLE1  'JOHN DOE';
RUN;
QUIT;
```

The SAS Output

```
                        JOHN DOE                              1

                   The GLM Procedure

                Class Level Information

        Class          Levels    Values

        NUM_EXP             3    0_EXP 2_EXP 4_EXP

                Number of observations    21
```

JOHN DOE 2

The GLM Procedure

Dependent Variable: CONFID

Source	DF	Sum of Squares	Mean Square	F Value	Pr > F
Model	2	29.18452381	14.59226190	6.97	0.0057
Error	18	37.67857143	2.09325397		
Corrected Total	20	66.86309524			

R-Square	Coeff Var	Root MSE	CONFID Mean
0.436482	35.95618	1.446808	4.023810

Source	DF	Type I SS	Mean Square	F Value	Pr > F
NUM_EXP	2	29.18452381	14.59226190	6.97	0.0057

Source	DF	Type III SS	Mean Square	F Value	Pr > F
NUM_EXP	2	29.18452381	14.59226190	6.97	0.0057

```
                         JOHN DOE                                  3

                    The GLM Procedure

      Level of             ------------CONFID-----------
      NUM_EXP       N              Mean              Std Dev

      0_EXP         7        2.35714286           1.23201345
      2_EXP         7        4.82142857           1.57925540
      4_EXP         7        4.89285714           1.50594062


                         JOHN DOE                                  4

                    The GLM Procedure

     Tukey's Studentized Range (HSD) Test for CONFID

  NOTE: This test controls the Type I experimentwise error rate.


       Alpha                                          0.05
       Error Degrees of Freedom                         18
       Error Mean Square                          2.093254
       Critical Value of Studentized Range         3.60930
       Minimum Significant Difference               1.9737


 Comparisons significant at the 0.05 level are indicated by ***.


                        Difference
     NUM_EXP               Between     Simultaneous 95%
   Comparison                Means    Confidence Limits

   4_EXP - 2_EXP            0.0714     -1.9023   2.0451
   4_EXP - 0_EXP            2.5357      0.5620   4.5094   ***
   2_EXP - 4_EXP           -0.0714     -2.0451   1.9023
   2_EXP - 0_EXP            2.4643      0.4906   4.4380   ***
   0_EXP - 4_EXP           -2.5357     -4.5094  -0.5620   ***
   0_EXP - 2_EXP           -2.4643     -4.4380  -0.4906   ***
```

ANOVA Summary Table

Table 15.S1.1

ANOVA Summary Table for Study Investigating the Relationship Between Exposure to Suggestions and the Formation of False Memories

Source	df	SS	MS	F	R^2
Number of exposures	2	29.18	14.59	6.97 *	.44
Within groups	18	37.68	2.09		
Total	20	66.86			

Note: N = 21
* p < .01

Confidence Intervals for Differences Between Means

Table 15.S1.2

Results of Tukey Tests Comparing (A) Zero-Exposure Group, (B) Two-Exposure Group, and (C) Four-Exposure Group on the Criterion Variable (Confidence that the Thief Wore Gloves)

Comparison [a]	Difference between means	Simultaneous 95% confidence limits: Lower	Upper
C - B	0.071	-1.902	2.045
C - A	2.536 *	0.562	4.509
B - A	2.464 *	0.491	4.438

Note. N = 21.
[a] Differences are computed by subtracting the mean for the second group from the mean for the first group.
* Tukey test indicates that the differences between the means is significant at p < .05.

Report Summarizing the Results of the Analysis

A) <u>Statement of the research question</u>: The purpose of this study was to determine whether there was a relationship between (a) the number of exposures to misleading suggestions made during questioning and (b) subject confidence that the thief in the video wore gloves.

B) <u>Statement of the research hypothesis</u>: There will be a positive relationship between the number of exposures to misleading suggestions and subject confidence that the thief wore gloves. Specifically, it is predicted that (a) subjects exposed to four suggestions will demonstrate a higher level of confidence than subjects exposed to two or zero suggestions, and (b) subjects exposed to two suggestions will demonstrate a higher level of confidence than subjects exposed to zero suggestions.

C) <u>Nature of the variables</u>: This analysis involved one predictor variable and one criterion variable:

- The predictor variable was the number of exposures to misleading suggestions. This was a limited-value variable, was assessed on a ratio scale, and included three levels: zero exposures, two exposures, and four exposures.
- The criterion variable was subjects' rated confidence that the thief wore gloves. This was a multi-value variable, and was assessed on an interval scale.

D) <u>Statistical test</u>: One-way ANOVA with one between-subjects factor.

E) <u>Statistical null hypothesis</u> (H_0): $\mu_1 = \mu_2 = \mu_3$; In the study population, there is no difference between subjects in the zero-exposures condition, subjects in the two-exposures condition, and subjects in the four-exposures condition with respect to their mean scores for the criterion variable (confidence that the thief wore gloves).

F) Statistical alternative hypothesis (H_1): Not all μs are equal; In the study population, there is a difference between at least two of the following three groups with respect to their mean scores for the confidence criterion variable: subjects in the zero-exposures condition, subjects in the two-exposures condition, and subjects in the four-exposures condition.

G) Obtained statistic: $F(2, 18) = 6.97$

H) Obtained probability (p) value: $p = .0057$

I) Conclusion regarding the statistical null hypothesis: Reject the null hypothesis.

J) Multiple comparison procedure: Tukey's HSD test showed that subjects in the four-exposures condition and two-exposures condition scored significantly higher on confidence than did subjects in the zero-exposures condition ($p < .05$). With alpha set at .05, there were no significant differences between subjects in the four-exposures condition and those in the two-exposures condition.

K) Confidence intervals: Confidence intervals for differences between the means are presented in Table 15.S1.2.

L) Effect size: $R^2 = .44$, indicating that the number of misleading suggestions accounted for 44% of the variance in subject confidence that the thief wore gloves.

M) Conclusion regarding the research hypothesis: These findings provide partial support for the study's research hypothesis. The findings provided support for the hypothesis that (a) subjects exposed to four suggestions will demonstrate a higher level of confidence than subjects exposed to zero suggestions, as well as for the hypothesis that (b) subjects exposed to two suggestions will demonstrate a higher level of confidence than subjects exposed to zero suggestions. However, the study failed to provide support for the hypothesis that subjects exposed to four suggestions will demonstrate a higher level of confidence than subjects exposed to two suggestions.

N) Formal description of the results for a paper:

Results were analyzed using a one-way ANOVA with one between-subjects factor. This analysis revealed a significant treatment effect for the number of exposures to misleading suggestions, $F(2, 18) = 6.97$, $MSE = 2.09$, $p = .0057$.

For the criterion variable (rated confidence that the thief wore gloves), the mean score for the four-exposures condition was 4.89 ($SD = 1.51$), the mean score for the two-exposures condition was 4.82 ($SD = 1.58$), and the mean for the zero-exposures condition was 2.36 ($SD = 1.23$). The sample means are displayed in Figure 15.S1.1. Tukey's HSD test showed that subjects in the four-exposures condition and two-exposures condition scored significantly higher on confidence than did subjects in the zero-exposures condition ($p < .05$). With alpha set at .05, there were no significant differences between subjects in the four-exposures condition and those in the two-exposures condition. Confidence intervals for differences between the means are presented in Table 15.S1.2.

In the analysis, R^2 was computed as .44. This indicated that the number of misleading suggestions accounted for 44 percent of the variance in subject confidence that the thief wore gloves.

O) Figure representing the results: See Figure 15.S1.1.

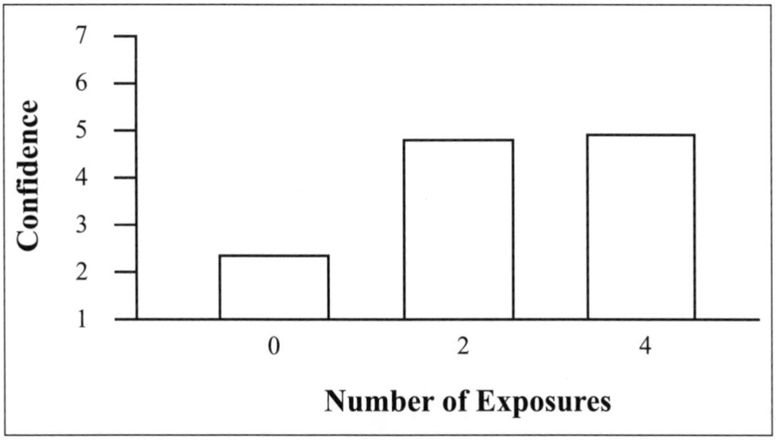

Figure 15.S1.1: Mean level of confidence as a function of the number of exposures to misleading suggestions.

Solution for Chapter 16: Factorial ANOVA with Two Between-Subjects Factors

16 SOLUTION

Exercise 16.1: The Effects of Misleading Suggestions and Pre-Event Instructions on the Creation of False Memories

The SAS Program

```
OPTIONS LS=80  PS=60;
DATA D1;
   INPUT   SUB_NUM
           INSTRUCT  $
           NUM_EXP   $
           CONFID       ;
DATALINES;
01  W  0_EXP 1.50
02  W  0_EXP 1.25
03  W  0_EXP 2.00
04  W  0_EXP 1.75
05  W  2_EXP 1.50
06  W  2_EXP 2.00
07  W  2_EXP 1.25
08  W  2_EXP 1.50
09  W  4_EXP 1.50
10  W  4_EXP 1.75
11  W  4_EXP 2.00
12  W  4_EXP 1.50
13  N  0_EXP 2.50
```

Continued on the next page

Continued from the previous page

```
14  N  0_EXP 2.50
15  N  0_EXP 2.00
16  N  0_EXP 3.00
17  N  2_EXP 5.00
18  N  2_EXP 5.50
19  N  2_EXP 4.00
20  N  2_EXP 4.75
21  N  4_EXP 5.50
22  N  4_EXP 5.75
23  N  4_EXP 5.75
24  N  4_EXP 5.25
;

PROC GLM  DATA=D1;
   CLASS  INSTRUCT  NUM_EXP;
   MODEL  CONFID = INSTRUCT  NUM_EXP
          INSTRUCT*NUM_EXP;
   MEANS  INSTRUCT  NUM_EXP  INSTRUCT*NUM_EXP;
   MEANS  INSTRUCT  NUM_EXP /  TUKEY  CLDIFF
          ALPHA=0.05;
   TITLE1 'JANE DOE';
RUN;
QUIT;
```

The SAS Output

```
                         JANE DOE                                   1

                     The GLM Procedure

                  Class Level Information

              Class         Levels    Values

              INSTRUCT           2    N W

              NUM_EXP            3    0_EXP 2_EXP 4_EXP

                 Number of observations    24
```

```
                       JANE DOE                                2

                  The GLM Procedure

Dependent Variable: CONFID

                              Sum of
 Source              DF      Squares    Mean Square   F Value   Pr > F
 Model                5  63.08333333    12.61666667     86.51   <.0001
 Error               18   2.62500000     0.14583333
 Corrected Total     23  65.70833333


          R-Square     Coeff Var      Root MSE    CONFID Mean
          0.960051      12.90866      0.381881       2.958333


 Source              DF    Type I SS    Mean Square   F Value   Pr > F
 INSTRUCT             1  42.66666667    42.66666667    292.57   <.0001
 NUM_EXP              2  10.39583333     5.19791667     35.64   <.0001
 INSTRUCT*NUM_EXP     2  10.02083333     5.01041667     34.36   <.0001


 Source              DF  Type III SS    Mean Square   F Value   Pr > F
 INSTRUCT             1  42.66666667    42.66666667    292.57   <.0001
 NUM_EXP              2  10.39583333     5.19791667     35.64   <.0001
 INSTRUCT*NUM_EXP     2  10.02083333     5.01041667     34.36   <.0001
```

```
                        JANE DOE                                    3

                   The GLM Procedure

       Level of                  ------------CONFID-----------
       INSTRUCT      N                  Mean           Std Dev

       N            12            4.29166667        1.42156018
       W            12            1.62500000        0.27177865


        Level of                 ------------CONFID-----------
        NUM_EXP      N                  Mean           Std Dev

        0_EXP        8            2.06250000        0.57863756
        2_EXP        8            3.18750000        1.79657412
        4_EXP        8            3.62500000        2.08309522


  Level of     Level of              ------------CONFID-----------
  INSTRUCT     NUM_EXP      N               Mean           Std Dev

  N            0_EXP        4         2.50000000        0.40824829
  N            2_EXP        4         4.81250000        0.62500000
  N            4_EXP        4         5.56250000        0.23935678
  W            0_EXP        4         1.62500000        0.32274861
  W            2_EXP        4         1.56250000        0.31457643
  W            4_EXP        4         1.68750000        0.23935678
```

```
                    JANE DOE                               4

               The GLM Procedure

      Tukey's Studentized Range (HSD) Test for CONFID

NOTE: This test controls the Type I experiment-wise error rate.


        Alpha                                        0.05
        Error Degrees of Freedom                       18
        Error Mean Square                        0.145833
        Critical Value of Studentized Range       2.97115
        Minimum Significant Difference             0.3275


Comparisons significant at the 0.05 level are indicated by ***.


                    Difference
      INSTRUCT         Between     Simultaneous 95%
     Comparison          Means    Confidence Limits

     N     - W          2.6667      2.3391   2.9942  ***
     W     - N         -2.6667     -2.9942  -2.3391  ***
```

```
                              JANE DOE                                5

                          The GLM Procedure

          Tukey's Studentized Range (HSD) Test for CONFID

   NOTE: This test controls the Type I experiment-wise error rate.


             Alpha                                       0.05
             Error Degrees of Freedom                      18
             Error Mean Square                       0.145833
             Critical Value of Studentized Range      3.60930
             Minimum Significant Difference            0.4873


   Comparisons significant at the 0.05 level are indicated by ***.


                           Difference
             NUM_EXP          Between     Simultaneous 95%
           Comparison           Means    Confidence Limits

          4_EXP - 2_EXP        0.4375     -0.0498   0.9248
          4_EXP - 0_EXP        1.5625      1.0752   2.0498  ***
          2_EXP - 4_EXP       -0.4375     -0.9248   0.0498
          2_EXP - 0_EXP        1.1250      0.6377   1.6123  ***
          0_EXP - 4_EXP       -1.5625     -2.0498  -1.0752  ***
          0_EXP - 2_EXP       -1.1250     -1.6123  -0.6377  ***
```

ANOVA Summary Table

Table 16.S1.1

ANOVA Summary Table for Study Investigating the Relationship Between Exposure to Suggestions (A), Pre-Event Instructions (B), and the Formation of False Memories

Source	df	SS	MS	F	p	R^2
Number of exposures (A)	2	10.40	5.20	35.64	.0001	.16
Pre-event instructions (B)	1	42.67	42.67	292.57	.0001	.65
A X B Interaction	2	10.02	5.01	34.36	.0001	.15
Within groups	18	2.63	0.15			
Total	23	65.71				

Note: N = 24.

Analysis Report Concerning the Interaction

A) Statement of the research question: The purpose of this study was to determine whether there was a significant interaction between (a) the number of exposures to misleading suggestions and (b) pre-event instructions in the prediction of (c) subject confidence that the thief in the video wore gloves.

B) Statement of the research hypothesis: The positive relationship between number of exposures and subject confidence will be stronger for non-warned subjects than for warned subjects.

C) Nature of the variables: This analysis involved two predictor variables and one criterion variable:

- Predictor A was the number of exposures to misleading suggestions. This was a limited-value variable, was assessed on an ratio scale, and included three levels: zero exposures, two exposures, and four exposures.

- Predictor B was the pre-event instructions. This was a dichotomous variable, was assessed on a nominal scale, and included two levels: warned and non-warned.
- The criterion variable was the rated confidence that the thief wore gloves. This was a multi-value variable, and was assessed on an interval scale.

D) Statistical test: Factorial ANOVA with two between-subjects factors.

E) Statistical null hypothesis (H_0): In the population, there is no interaction between the number of exposures to misleading suggestions and pre-event instructions in the prediction of the criterion variable (subject confidence that the thief in the video wore gloves).

F) Statistical alternative hypothesis (H_1): In the population, there is an interaction between the number of exposures to misleading suggestions and pre-event instructions in the prediction of the criterion variable (subject confidence that the thief in the video wore gloves).

G) Obtained statistic: $F(2, 18) = 34.36$

H) Obtained probability (p) value: $p = .0001$

I) Conclusion regarding the statistical null hypothesis: Reject the null hypothesis.

J) Multiple comparison procedure. Not relevant.

K) Confidence intervals. Not relevant.

L) Effect size. $R^2 = .15$, indicating that the interaction term accounted for 15 percent of the variance in subject confidence that that the thief wore gloves.

M) <u>Conclusion regarding the research hypothesis</u>: These findings provide support for the study's research hypothesis that the positive relationship between number of exposures and subject confidence will be stronger for non-warned subjects than for warned subjects.

N) <u>Formal description of the results for a paper</u>:

Results were analyzed using a factorial ANOVA with two between-subjects factors. This analysis revealed a significant F statistic for the interaction between number of exposures and pre-event instructions, $F(2, 18) = 34.36$, $MSE = 0.15$ $p = .0001$.

Sample means for the various conditions that constituted the study are displayed in Figure 16.S1.1. The nature of the interaction displayed in Figure 16.S1.1 shows that there is a positive relationship between number of exposures and confidence for subjects in the non-warned group: for subjects in this condition, a greater number of misleading suggestions was associated with greater confidence that the thief wore gloves. On the other hand, there was only a very weak relationship between number of exposures and confidence for subjects in the warned group.

In the analysis, R^2 for this interaction effect was computed as .15. This indicated that the interaction accounted for 15 percent of the variance in subject confidence that the thief wore gloves.

O) <u>Figure representing the results</u>: See figure 16.S1.1.

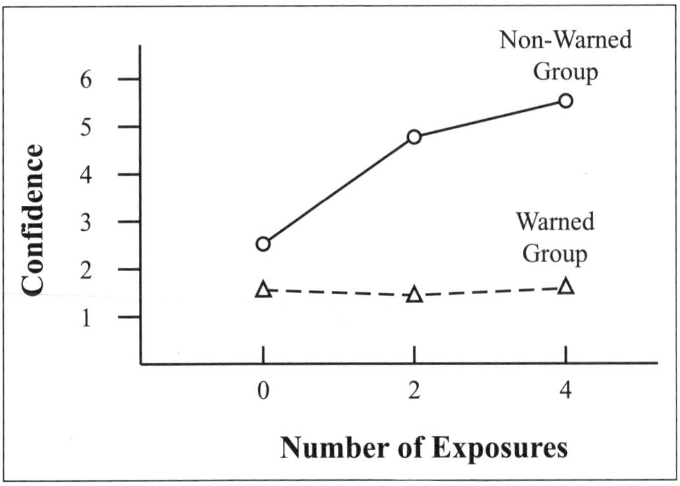

Figure 16.S1.1. Mean level of confidence as a function of the number of exposures to misleading suggestions and pre-event instructions.

Solution for Chapter 17: Chi-Square Test of Independence

17
SOLUTION

Solution to Exercise 17.1: The Relationship Between Sex of Children and Marital Disruption

The SAS Program

```
OPTIONS  LS=80  PS=60;
DATA D1;
   INPUT  STATUS    $
          SEX       $
          NUMBER;
DATALINES;
I  BB  34
I  BG  26
I  GG  15
S  BB  14
S  BG  22
S  GG  36
;
PROC FREQ  DATA=D1;
   TABLES  STATUS*SEX  /  ALL;
   WEIGHT  NUMBER;
   TITLE1 'JOHN DOE';
RUN;
```

The SAS Output

```
                      JOHN DOE                                  1

                 The FREQ Procedure

               Table of STATUS by SEX

        STATUS     SEX

        Frequency|
        Percent  |
        Row Pct  |
        Col Pct  |BB      |BG      |GG      |  Total
        ---------+--------+--------+--------+
        I        |     34 |     26 |     15 |     75
                 |  23.13 |  17.69 |  10.20 |  51.02
                 |  45.33 |  34.67 |  20.00 |
                 |  70.83 |  54.17 |  29.41 |
        ---------+--------+--------+--------+
        S        |     14 |     22 |     36 |     72
                 |   9.52 |  14.97 |  24.49 |  48.98
                 |  19.44 |  30.56 |  50.00 |
                 |  29.17 |  45.83 |  70.59 |
        ---------+--------+--------+--------+
        Total          48       48       51      147
                    32.65    32.65    34.69   100.00


          Statistics for Table of STATUS by SEX

   Statistic                     DF       Value      Prob
   ------------------------------------------------------
   Chi-Square                     2     17.2597    0.0002
   Likelihood Ratio Chi-Square    2     17.7751    0.0001
   Mantel-Haenszel Chi-Square     1     16.9322    <.0001
   Phi Coefficient                       0.3427
   Contingency Coefficient               0.3242
   Cramer's V                            0.3427
```

JOHN DOE 2

The FREQ Procedure

Statistics for Table of STATUS by SEX

Statistic	Value	ASE
Gamma	0.5257	0.1037
Kendall's Tau-b	0.3214	0.0714
Stuart's Tau-c	0.3710	0.0824
Somers' D C|R	0.3711	0.0824
Somers' D R|C	0.2783	0.0619
Pearson Correlation	0.3405	0.0757
Spearman Correlation	0.3409	0.0757
Lambda Asymmetric C|R	0.1979	0.0653
Lambda Asymmetric R|C	0.2917	0.0835
Lambda Symmetric	0.2381	0.0625
Uncertainty Coefficient C|R	0.0551	0.0251
Uncertainty Coefficient R|C	0.0873	0.0398
Uncertainty Coefficient Symmetric	0.0675	0.0308

Sample Size = 147

Summary Statistics for STATUS by SEX

Cochran-Mantel-Haenszel Statistics (Based on Table Scores)

Statistic	Alternative Hypothesis	DF	Value	Prob
1	Nonzero Correlation	1	16.9322	<.0001
2	Row Mean Scores Differ	1	16.9322	<.0001
3	General Association	2	17.1423	0.0002

Total Sample Size = 147

Analysis Report

A) <u>Statement of the research question</u>: The purpose of this study was to determine whether there was a relationship between (a) the sex of children and (b) marital status (intact versus separated) at the end of a 15-year interval. Specifically, this study was designed to determine whether there was a difference between families with two boys, families with a boy and a girl, and families with two girls with respect to the parents' marital status (intact versus separated) at the end of a 15-year interval.

B) <u>Statement of the research hypothesis</u>: There will be a relationship between the sex of children and marital status such that (a) a larger percentage of families that have two boys will be intact, compared to families with a boy and a girl or families with two girls, and (b) a larger percentage of families that have a boy and a girl will be intact, compared to families with two girls.

C) <u>Nature of the variables</u>: This analysis involved two variables:

- The predictor variable was sex of children. This was a limited-value variable, was measured on a nominal scale and could have three values: two boys, a boy and a girl, and two girls.

- The criterion variable was marital status. This was a dichotomous variable, was measured on a nominal scale and could have two values: intact and separated.

D) <u>Statistical test</u>: Chi-square test of independence.

E) <u>Statistical null hypothesis</u> (H_0): In the study population, there is no relationship between the sex of children and marital status.

F) <u>Statistical alternative hypothesis</u> (H_1): In the study population, there is a relationship between the sex of children and marital status.

G) <u>Obtained statistic</u>: χ^2 (2, $\underline{N}$ = 147) = 17.260.

H) <u>Obtained probability (p) value</u>: $\underline{p}$ = .0002.

I) <u>Conclusion regarding the statistical null hypothesis</u>: Reject the null hypothesis.

J) <u>Effect size</u>: Cramer's $\underline{V}$ was used as the index of effect size. For this analysis, Cramer's $\underline{V}$ = .34. Values of Cramer's $\underline{V}$ may range from zero to +1.00, with values closer to zero indicating a weaker relationship between the predictor variable and the criterion variable.

K) <u>Conclusion regarding the research hypothesis</u>: These findings provide support for the study's research hypothesis.

L) <u>Formal description of the results for a paper</u>:

Results were analyzed using a chi-square test of independence. This analysis revealed a significant relationship between the sex of children and marital status, χ^2 (2, $\underline{N}$ = 147) = 17.260, $\underline{p}$ = .0002. Figure 17.S1.1 illustrates the number of families that were intact versus separated, broken down according to the sex of the children. The crosstabulation table showed that, for families with two boys, the percentage of couples who were intact was much larger than the percentage of those who were separated (71 percent versus 29 percent, respectively); for families with a boy and a girl, the percentage of couples who were intact was slightly larger than the percentage of those who were separated (54 percent versus 46 percent, respectively); and for families with two girls, the percentage of those who were intact was smaller that the percentage of those who were separated (29 percent versus 71 percent, respectively).

Cramer's V was used to assess the strength of the relationship between the two variables. This statistic may range from zero to +1.00, with values closer to zero indicating a weaker relationship. For this analysis, Cramer's V was computed as V = .34.

M) Figure representing the results. See Figure 17.S1.1.

Figure Representing the Results

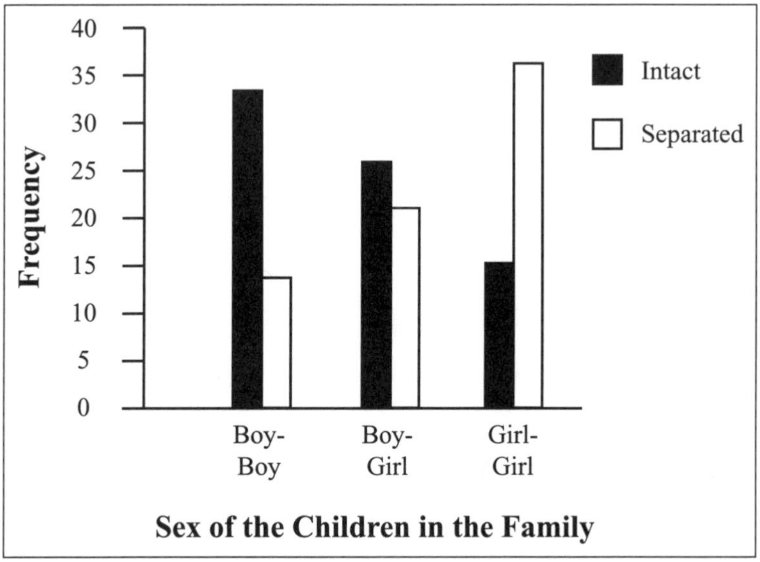

Figure 17.S1.1. Marital status of parents (intact versus separated) as a function of the sex of the children in the family.

This combined index includes entries for Basic Statistics Using SAS: Student Guide *and* Exercises.

Page numbers preceded by "E" indicate pages in this book. All other page numbers refer to Basic Statistics Using SAS: Student Guide.

Index

A

D

E

G

H

I

J

K

L

M

N

O

P

Q

R

T

U

V

W

Y

Z

Special Characters

Call your local SAS office to order these books from Books by Users Press

Advanced Log-Linear Models Using SAS®
by **Daniel Zelterman** . . . . Order No. A57496

Annotate: Simply the Basics
by **Art Carpenter** . . . . Order No. A57320

Applied Multivariate Statistics with SAS® Software, Second Edition
by **Ravindra Khattree**
and **Dayanand N. Naik** . . . . Order No. A56903

Applied Statistics and the SAS® Programming Language, Fourth Edition
by **Ronald P. Cody**
and **Jeffrey K. Smith** . . . . Order No. A55984

An Array of Challenges — Test Your SAS® Skills
by **Robert Virgile** . . . . Order No. A55625

Beyond the Obvious with SAS® Screen Control Language
by **Don Stanley** . . . . Order No. A55073

Carpenter's Complete Guide to the SAS® Macro Language
by **Art Carpenter** . . . . Order No. A56100

The Cartoon Guide to Statistics
by **Larry Gonick**
and **Woollcott Smith** . . . . Order No. A5515

Categorical Data Analysis Using the SAS® System, Second Edition
by **Maura E. Stokes, Charles S. Davis,**
and **Gary G. Koch** . . . . Order No. A57998

Cody's Data Cleaning Techniques Using SAS® Software
by **Ron Cody** . . . . Order No. A57198

Common Statistical Methods for Clinical Research with SAS® Examples, Second Edition
by **Glenn A. Walker** . . . . Order No. A58086

Concepts and Case Studies in Data Management
by **William S. Calvert**
and **J. Meimei Ma** . . . . Order No. A55220

Debugging SAS® Programs: A Handbook of Tools and Techniques
by **Michele M. Burlew** . . . . Order No. A57743

Efficiency: Improving the Performance of Your SAS® Applications
by **Robert Virgile** . . . . Order No. A55960

A Handbook of Statistical Analyses Using SAS®, Second Edition
by **B.S. Everitt**
and **G. Der** . . . . Order No. A58679

Health Care Data and the SAS® System
by **Marge Scerbo, Craig Dickstein,**
and **Alan Wilson** . . . . Order No. A57638

The How-To Book for SAS/GRAPH® Software
by **Thomas Miron** . . . . Order No. A55203

In the Know... SAS® Tips and Techniques From Around the Globe
by **Phil Mason** . . . . Order No. A55513

Integrating Results through Meta-Analytic Review Using SAS® Software
by **Morgan C. Wang**
and **Brad J. Bushman** .... Order No. A55810

Learning SAS® in the Computer Lab, Second Edition
by **Rebecca J. Elliott** .... Order No. A57739

The Little SAS® Book: A Primer
by **Lora D. Delwiche**
and **Susan J. Slaughter** .... Order No. A55200

The Little SAS® Book: A Primer, Second Edition
by **Lora D. Delwiche**
and **Susan J. Slaughter** .... Order No. A56649
(updated to include Version 7 features)

Logistic Regression Using the SAS® System: Theory and Application
by **Paul D. Allison** .... Order No. A55770

Longitudinal Data and SAS®: A Programmer's Guide
by **Ron Cody** .... Order No. A58176

Maps Made Easy Using SAS®
by **Mike Zdeb** .... Order No. A57495

Models for Discrete Date
by **Daniel Zelterman** .... Order No. A57521

Multiple Comparisons and Multiple Tests Using SAS® Text and Workbook Set
(books in this set also sold separately)
by **Peter H. Westfall, Randall D. Tobias, Dror Rom, Russell D. Wolfinger**
and **Yosef Hochberg** .... Order No. A55770

Multiple-Plot Displays: Simplified with Macros
by **Perry Watts** .... Order No. A58314

Multivariate Data Reduction and Discrimination with SAS® Software
by **Ravindra Khattree,**
and **Dayanand N. Naik** .... Order No. A56902

The Next Step: Integrating the Software Life Cycle with SAS® Programming
by **Paul Gill** .... Order No. A55697

Output Delivery System: The Basics
by **Lauren E. Haworth** .... Order No. A58087

Painless Windows: A Handbook for SAS® Users
by **Jodie Gilmore** .... Order No. A55769
(for Windows NT and Windows 95)

Painless Windows: A Handbook for SAS® Users, Second Edition
by **Jodie Gilmore** .... Order No. A56647
(updated to include Version 7 features)

PROC TABULATE by Example
by **Lauren E. Haworth** .... Order No. A56514

Professional SAS® Programmer's Pocket Reference, Fourth Edition
by **Rick Aster** .... Order No. A58128

Professional SAS® Programmer's Pocket Reference, Second Edition
by **Rick Aster** .... Order No. A56646

Professional SAS® Programming Shortcuts
by **Rick Aster** .... Order No. A59353

Programming Techniques for Object-Based Statistical Analysis with SAS® Software
by **Tanya Kolosova**
and **Samuel Berestizhevsky** .... Order No. A55869

Quick Results with SAS/GRAPH® Software
by **Arthur L. Carpenter**
and **Charles E. Shipp** .... Order No. A55127

Quick Results with the Output Delivery System
by **Sunil Gupta** .... Order No. A58458

Quick Start to Data Analysis with SAS®
by **Frank C. Dilorio**
and **Kenneth A. Hardy** .... Order No. A55550

Reading External Data Files Using SAS®: Examples Handbook
by **Michele M. Burlew** .... Order No. A58369

*Welcome * Bienvenue * Willkommen * Yohkoso * Bienvenido*

SAS Publishing Is Easy to Reach

'isit our Web site located at support.sas.com/pubs

ɔu will find product and service details, including

- **companion Web sites**
- **sample chapters**
- **tables of contents**
- **author biographies**
- **book reviews**

earn about

- **regional users group conferences**
- **trade show sites and dates**
- **authoring opportunities**
- **e-books**

:xplore all the services that Publications has to offer!

'our Listserv Subscription Automatically Brings the News to You

o you want to be among the first to learn about the latest books and services available from SAS ublishing? Subscribe to our listserv **newdocnews-l** and, once each month, you will automatically :ceive a description of the newest books and which environments or operating systems and SAS® :lease(s) each book addresses.

ɔ subscribe,

. Send an e-mail message to **listserv@vm.sas.com**.

. Leave the "Subject" line blank.

. Use the following text for your message:

   subscribe NEWDOCNEWS-L ***your-first-name your-last-name***

   For example: subscribe NEWDOCNEWS-L John Doe

You're Invited to Publish with SAS Institute's Books by Users Press

If you enjoy writing about SAS software and how to use it, the Books by Users program at SAS Institute offers a variety of publishing options. We are actively recruiting authors to publish book and sample code.

If you find the idea of writing a book by yourself a little intimidating, consider writing with a co-author. Keep in mind that you will receive complete editorial and publishing support, access t our users, technical advice and assistance, and competitive royalties. Please ask us for an author packet at **sasbbu@sas.com** or call 919-531-7447. See the Books by Users Web page at **support.sas.com/bbu** for complete information.

Book Discount Offered at SAS Public Training Courses!

When you attend one of our SAS Public Training Courses at any of our regional Training Cente in the United States, you will receive a 20% discount on any book orders placed during the course. Take advantage of this offer at the next course you attend!

SAS Institute Inc.
SAS Campus Drive
Cary, NC 27513-2414
Fax 919-677-4444

E-mail: sasbook@sas.com
Web page: support.sas.com/pubs
To order books, call SAS Publishing Sales at 800-727-3228*
For product information, consulting, customer service, or training, call 800-727-0025
For other SAS Institute business, call 919-677-8000*

* **Note:** Customers outside the United States should contact their local SAS office.